Lecture Notes in Mobility

The book series Lecture Notes in Mobility (LNMOB) reports on innovative, peer-reviewed research and developments in intelligent, connected and sustainable transportation systems of the future. It covers technological advances, research, developments and applications, as well as business models, management systems and policy implementation relating to: zero-emission, electric and energy-efficient vehicles; alternative and optimized powertrains; vehicle automation and cooperation; clean, user-centric and on-demand transport systems; shared mobility services and intermodal hubs; energy, data and communication infrastructure for transportation; and micromobility and soft urban modes, among other topics. The series gives a special emphasis to sustainable, seamless and inclusive transformation strategies and covers both traditional and any new transportation modes for passengers and goods. Cutting-edge findings from public research funding programs in Europe, America and Asia do represent an important source of content for this series. PhD thesis of exceptional value may also be considered for publication. Supervised by a scientific advisory board of world-leading scholars and professionals, the Lecture Notes in Mobility are intended to offer an authoritative and comprehensive source of information on the latest transportation technology and mobility trends to an audience of researchers, practitioners, policymakers, and advanced-level students, and a multidisciplinary platform fostering the exchange of ideas and collaboration between the different groups.

Shuai Su · Zhongbei Tian · Rob M. P. Goverde

Energy-Efficient Train Operation

A System Approach for Railway Networks

Springer

Shuai Su
State Key Laboratory of Rail Traffic
Control and Safety
Beijing Jiaotong University
Beijing, China

Zhongbei Tian
School of Engineering
University of Birmingham
Birmingham, UK

Rob M. P. Goverde
Department of Transport and Planning
Delft University of Technology
Delft, The Netherlands

ISSN 2196-5544 ISSN 2196-5552 (electronic)
Lecture Notes in Mobility
ISBN 978-3-031-34655-2 ISBN 978-3-031-34656-9 (eBook)
https://doi.org/10.1007/978-3-031-34656-9

This Springer imprint is published by the registered company Springer Nature Switzerland AG
The registered company address is: Gewerbestrasse 11, 6330 Cham, Switzerland

Preface

Rail transport is developing rapidly across the world due to the merits of safety, convenience, and comfort. Although rail transport is environmentally friendly compared to other transport modes including road and air, the amount of energy consumption in the rail transport system can still be improved considering its large scale of operation and high frequency of train services. Reducing energy consumption also contributes to reducing energy costs and achieving decarbonisation. The railway sector is aiming at energy-saving measures in many fields, in which reducing the energy consumption of train operations gets much attention. For this reason, this book focuses on energy-efficient train operations. In this book, the principles and methods of energy-efficient train operation will be introduced to provide railway companies with a set of energy-saving methods that can be applied in practice. This book also provides a systematic introduction for researchers and students in the field of energy-efficient train operation, helping them to understand this field quickly and master the basic theoretical methods.

This book consists of eight chapters. Chapter 1 reviews energy consumption data and the main relevant projects in recent years. In addition, a classification is given for energy-efficient train operation research. Chapter 2 introduces the relationship between train operation and energy consumption under different energy-efficient strategies. This relationship will serve as the basis for the energy-efficient optimisation methods in the subsequent chapters. Four types of optimisation methods for energy-efficient train operation are proposed from Chaps. 3 to 6. In Chap. 3, the driving strategy optimisation method for a single train is introduced. Chapter 4 considers energy-efficient train timetabling for mainline railway corridors, including multiple stops and heterogeneous trains. Optimisation of timetables taking into consideration the regenerative braking energy for metro systems is presented in Chap. 5. To make full use of regenerative braking energy, Chap. 6 discusses the main technologies, modelling, and control methods of energy storage systems. Because traction power network modelling plays a significant role in validating energy-efficient train operations, Chap. 7 then presents the simulation of electric railway systems, which integrates the train movement model and railway power network model. Finally, basic conclusions about energy-efficient train operation methods and recommendations for further research are given in Chap. 8.

The main authors of this book are Shuai Su, Zhongbei Tian, and Rob Goverde. Professor Shuai Su is the deputy director of the Frontier Science Center of the Smart High-Speed Rail System at Beijing Jiaotong University. He has led projects concentrating on improving the energy efficiency of train operations. Moreover, he is a senior member of the Chinese Association of Automation and also a member of the TRB Annual Rail Transit Systems Committee. Dr. Zhongbei Tian has been working on railway traction power system modelling and analysis, energy-efficient train control, and energy system optimisation for more than ten years. He is now a Lecturer in Railway Power Systems at the University of Birmingham and an Honorary Lecturer at the University of Liverpool. His research has been implemented in projects across the world, including Network Rail, Edinburgh Tram in the UK, Madrid Metro in Spain, SMRT in Singapore, and Beijing and Guangzhou Metro in China. Professor Rob Goverde is a Professor of Railway Traffic Management and Operations and Director of the Digital Rail Traffic Lab at the Delft University of Technology. His research concentrates on the planning and management of railway traffic systems, including digitalised and automated train operations based on energy-efficient train trajectory optimization. He has wide experience from participation in many European railway projects and is President of the International Association of Railway Operations Research (IAROR) and Fellow of the Institution of Railway Signal Engineers.

The contributions of each author are listed as follows. Professor Shuai Su worked on Chaps. 1, 2, 5, and 8. Dr. Zhongbei Tian worked on Chaps. 1, 2, 7, and 8. Professor Rob Goverde worked on Chaps. 1, 2, 4, and 8. In addition, international experts were invited to contribute to chapters in the book. Xiao Liu is the co-author of Chap. 1. Xuekai Wang is the co-author of Chaps. 2 and 5. Peter Pudney is the author of Chap. 3. Gerben Scheepmaker is the co-author of Chap. 4. Gonzalo Sánchez-Contreras, Adrián Fernández Rodríguez, Antonio Fernández-Cardador, and Asunción Cucala are the authors of Chap. 6. We also like to thank all the support from students and experts of railway companies for this book.

It is, therefore, our pleasure to present this compendium. We hope that this book will support the railways all over the world to increase their contribution to clean mobility.

Beijing, China
Birmingham, UK
Delft, The Netherlands

Shuai Su
Zhongbei Tian
Rob M. P. Goverde

Contents

Chapter 1
Introduction to Energy-Efficient Train Operation

Zhongbei Tian, Xiao Liu, Shuai Su, and Rob M. P. Goverde

1.1 Background of Railway Energy Consumption

Energy and environmental sustainability in transportation have received increasing attention in recent decades. The Future of Rail—opportunities for energy and the environment, jointly published by The International Energy Agency (IEA) and the International Union of Railways (UIC) in 2019, underlined the global energy consumption data in the transport sector, particularly the railway [1]. On a global basis, the transport sector accounts for 29% of final energy use, and its energy demand has risen significantly in the past decade. The railway is one of the most energy-efficient modes of transport, which constituted 8% of passenger transport and 7% of freight movements globally in 2016 but only accounted for 2% of the energy used in the transport sector. Figure 1.1 shows the global final energy consumption in different sections [2].

From 1990 to 2015, even though the railway share of total transport activity kept above 8.5% (Fig. 1.2), the share of railway CO_2 emission was reduced to less than 3% (Fig. 1.3). Therefore, the railway plays an important role in reducing the environmental impact and improving energy efficiency. By offering efficient transport

Z. Tian (✉)
University of Birmingham, Birmingham, UK
e-mail: Z.Tian@bham.ac.uk

X. Liu
University of Liverpool, Liverpool, UK
e-mail: Xiao.Liu@liverpool.ac.uk

S. Su
Beijing Jiaotong University, Beijing, China
e-mail: shuaisu@bjtu.edu.cn

R. M. P. Goverde
Department of Transport and Planning, Delft University of Technology, Delft, The Netherlands
e-mail: R.M.P.Goverde@tudelft.nl

S. Su et al., *Energy-Efficient Train Operation*, Lecture Notes in Mobility,
https://doi.org/10.1007/978-3-031-34656-9_1

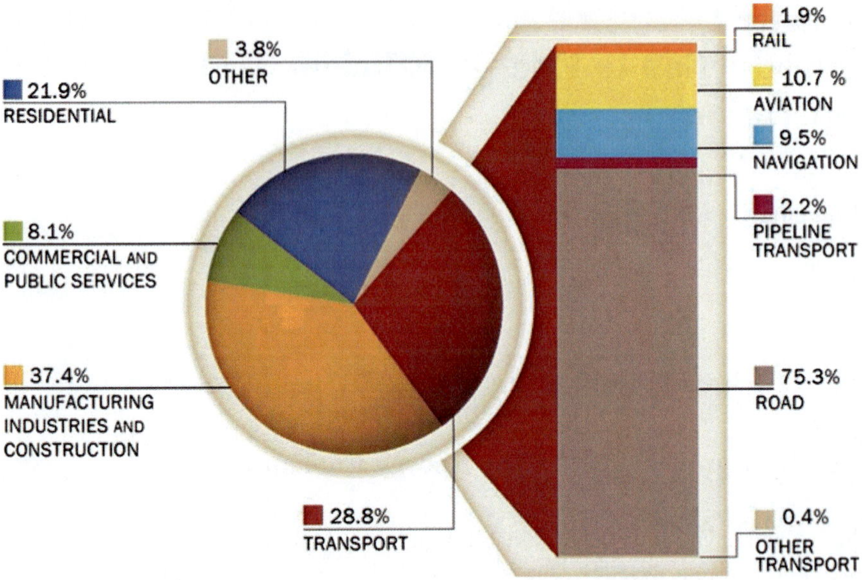

Fig. 1.1 Global energy consumption by sector [2]

with low environmental impacts, the railway helps create a more sustainable approach to transportation.

According to the various types of services, the railway can be classified into passenger and freight rail. The passenger rail includes urban rail, conventional rail, and high-speed rail. Urban rail transit generally refers to a railway system providing passenger services within metropolitan areas, which normally includes metro rail, light rail, and trams. Conventional rail normally serves medium- to long-distance train journeys with a maximum speed under 250 km/h and suburban train journeys connecting urban centres with surrounding areas. In terms of the rail infrastructure, conventional rail tracks refer to the tracks that can be used for passenger conventional rail and freight rail. High-speed rail is used for long-distance services which travel over 250 km/h. Figure 1.4 compares the track length for different rail types over the recent two decades. Conventional rail tracks account for 94% of all rail track-kilometres, but the length has grown slowly in recent decades. The high-speed rail track increases strongly in Europe and China. The Chinese high-speed rail expanded since 2005, and now accounts for nearly two-thirds of the world's high-speed rail lines. The urban rail lines increase gently in Europe and North America, but they expand significantly in Asia.

Hong Kong metro regularly transports 80,000 passengers per hour during peak time, which is four times higher than by bus [3]. In Tokyo, the share of public transport is 36%, while railway accounts for 91.7% of it in Fig. 1.5 [4]. Urban rail transit is also characterised by short headway and dwell time, and a high number of stations with short interstation distances. Urban rail systems can effectively satisfy

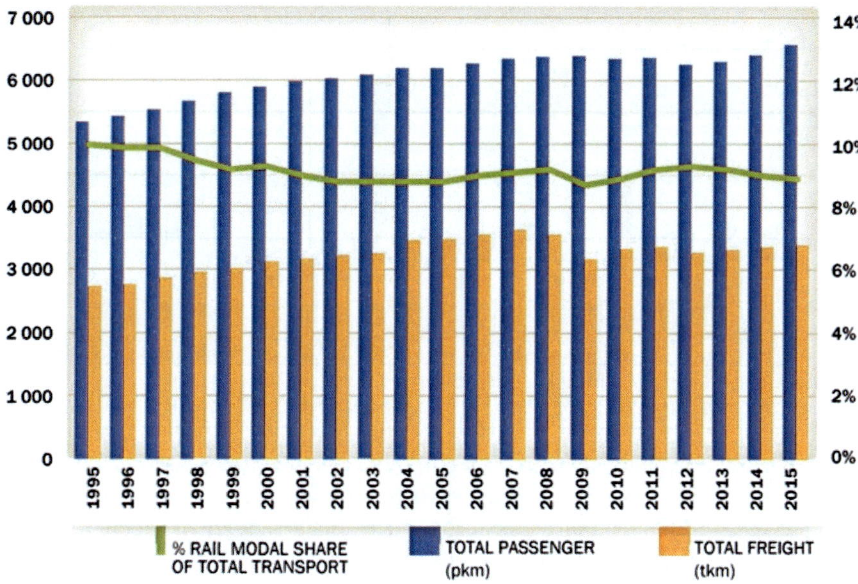

Fig. 1.2 Passenger and freight transport activity—all modes, 1995–2015 (billion pkm and tkm—left, share of rail over total—right) [2]

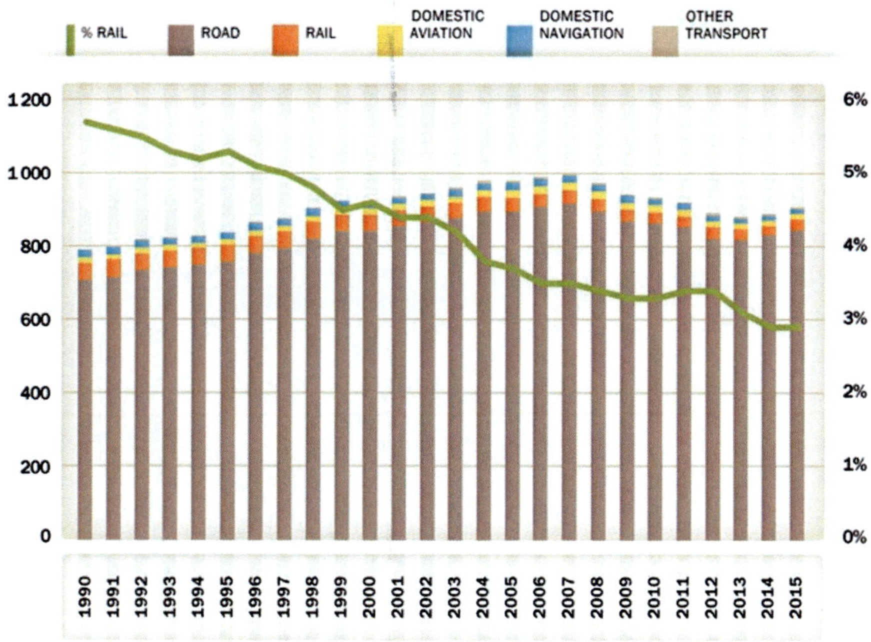

Fig. 1.3 Transport sector CO_2 emissions by mode, 1990–2015 (million t CO_2—left, rail share over total-right. Note: Electricity and heat production related emissions are reallocated to the end-use sectors.) [2]

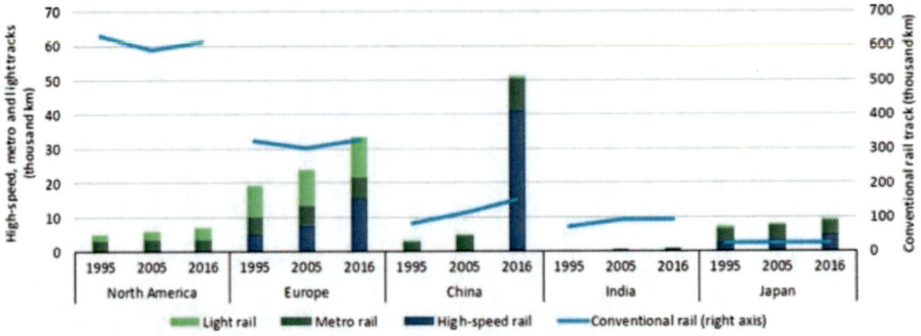

Fig. 1.4 Track length by region and network type, 1995–2016 (Note: Conventional rail includes infrastructure used both by conventional passenger and freight rail) [1]

high transportation demand and reduce air pollution in metropolitan areas. The total length of mainline railway, including passenger and freight, has also increased a lot in recent years. Significant investments have been made in high-speed rail in China, which has overtaken all other countries in terms of network length. Currently, about two-thirds of high-speed rail activity takes place in China [1]. The Chinese mainline is in charge of 22.9% of all passenger activity [5]. While in Australia, rail transport accounts for approximately 49% of total domestic freight, much higher than road freight, about 35%, and coastal sea freight 17% [6]. Additionally, freight rail offers the benefits of high throughput, low cost, and excellent safety while unaffected by the weather.

Fig. 1.5 Modal share in Tokyo

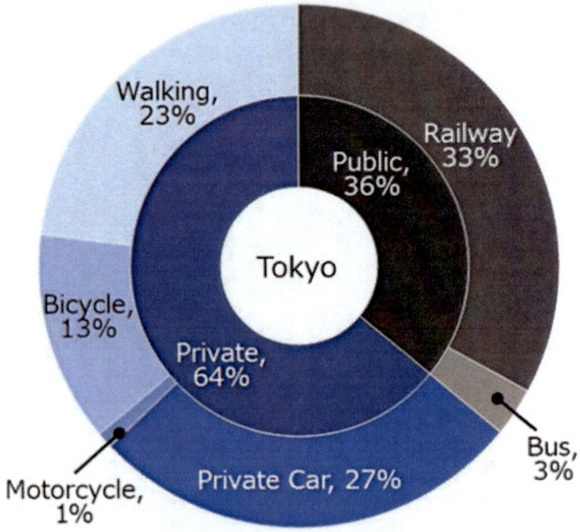

1.2 Projects in Railway Energy Efficiency

Although the railway system is arguably one of the most efficient forms of land-based transport, how to operate trains more efficiently is still of global importance. To improve sustainability, members of the International Union of Railways and the Community of European Railway and Infrastructure Companies proposed a unified approach to environmental and sustainability topics in the European rail sector in 2010 [7]. They addressed four targets for the rail sector to improve performance in terms of the environment, including climate protection, energy efficiency, exhaust emissions, and noise. European railway companies agreed to reduce specific average CO_2 emissions from train operation by 50% in 2030, compared to the emissions in 1990. In addition, it was agreed that by 2030 the energy consumption from train operation would be reduced by 30% compared to the consumption in 1990. There are a number of research and industry projects across the world studying energy efficiency in railways. This subsection provides an overview of the most relevant projects, including the national projects affiliated with the authors in this book.

1.2.1 European Projects

ON-TIME (Optimal Networks for Train Integration Management across Europe) was an EU project running during 2011–2014. The aim of the ON-TIME project was to develop new methods and processes to maximise the available capacity on the European railway network and to decrease overall delays in order to increase customer satisfaction and ensure that the railway network continues to provide a dependable, resilient, and green alternative to other modes of transport. ON-TIME proposed an integrated method to compute robust conflict-free train timetables including energy-efficient train trajectories, as a basis for efficient railway traffic management [8]. In addition, a real-time traffic management architecture was proposed in which traffic state monitoring and conflict detection and resolution modules were developed to maintain a Real-Time Traffic Plan (RTTP) consisting of conflict-free train paths and train orders over the railway network [9]. The RTTP is used both for automatic route setting and for calculating train path envelopes to feed Connected Driver Advisory Systems (C-DAS) to adopt an energy-efficient driving strategy. To allow the interoperable use of C-DAS throughout Europe, ON-TIME proposed a data format for communication of operational decisions (e.g., speed advice) between control centres and trains. Based on an extensive state-of-the-art analysis, three system architecture design alternatives and associated data formats were proposed to distribute the two key functions between trackside (control centre) and onboard components, i.e., (i) generating energy-efficient train speed profiles satisfying the targets and constraints of the train path envelope and (ii) presenting the corresponding advice to the driver [10–12]. The three DAS architecture alternatives and exchange data were further developed and standardized in the project SFERA (Smart communications for efficient rail activities) and are now part of the UIC code 90940 [13]. The train path envelope is here called the journey profile in analogy to the ATO-over-ETCS system requirement specification that is being developed for Automatic Train Operation running under the European Train Control System [14].

OPEUS (modelling and strategies for the assessment and OPtimisation of Energy USage aspects of rail innovation) is a project funded by Shift2Rail [15], which started in 2016 and ended in 2019. The aim of OPEUS was to develop a simulation methodology and an accompanying modelling tool to evaluate, improve and optimise the energy consumption of rail systems with a particular focus on in-vehicle innovation. The OPEUS concept was based on the need to understand and measure the energy being used by each of the relevant components of the rail system and, in particular, the vehicle, which includes the energy losses in the traction chain. New technologies were introduced to reduce these losses and optimise energy consumption. Specifically, the OPEUS approach had three components: (i) the energy simulation model, (ii) the energy use requirements (e.g., duty cycles), and (iii) the energy usage outlook and optimisation strategies recommendation. This project was built upon an extensive range of knowledge and outcomes generated by numerous key collaborative efforts. Significant complementary work from the academic community was used to enhance the activities of the project. With the help of simulation technology, this research created a programme that can calculate the energy consumption of various railway vehicles and their components. The energy KPI was also employed to quantify the relative savings of the technology demonstrator and summarise the overall savings per system platform demonstrator. Moreover, following the global trend, eco-labelling based on EN50591 was carried out to reinforce the attractiveness of railway transport.

In 2017, another Shift2Rail JU founded project, In2Stempo started [16]. This project focused on developing cost-efficient and reliable high-capacity infrastructure for the railway. The main components of this project included smart power supply, smart metering for Railway Distributed Energy Resource Management System, and future stations. The objectives of this project were to enhance the existing capacity fulfilling user demand for the European rail system, increase reliability by delivering better and consistent quality of service for the European rail system, and reduce the life cycle cost. These tasks were fulfilled in three steps: (i) develop a smart railway power grid in an interconnected and communicated system; (ii) achieve a fine mapping of energy flow within the entire railway system, forming the basis of later energy management strategy; (iii) improve the customer experience at Railway Stations. With the help of data analysis, process bus, digital twins, and machine learning, this project realised the smart power supply by developing a smart railway power grid in an interconnected and communicated system. Moreover, the fine mapping of energy flows and usage within the entire railway system was achieved by using smart metering. The outcomes of this project can help to build the future station by improving crowd management, station design, and accessibility to trains.

Apart from the research above, several other European projects have been carried out to improve the energy efficiency in railway systems. The Railenergy project, co-funded by the European Commission, started in 2006 to address the energy efficiency of the integrated railway system [17]. Recommendations included innovative traction technologies, components, and layouts for the development of rolling stock, operation, and infrastructure management strategies. The MERLIN project was conducted to investigate and demonstrate the viability of an integrated management system to achieve more sustainable and optimised energy usage in European electric mainline railway systems [18]. In 2012, 17 project partners collaborated on the OSIRIS project, including public transport operators, railway manufacturers, and research centres [19]. The OSIRIS project aimed to reduce energy consumption within European urban rail systems, focusing on developing and testing technological and operational solutions and tools.

1.2.2 Selected National Projects

In the UK, the Rail Safety and Standards Board (RSSB) is an independent safety, standards, and research body. They help to make an evolving railway safer, more efficient, and more sustainable. In 2017, RSSB launched a British rail research network, which focused on creating three centres of excellence, forming the heart of the British Railway Research and Innovation Network [20]. These centres include a digital system centre located at the University of Birmingham, a rolling stock centre led by the University of Huddersfield in collaboration with Newcastle and Loughborough universities, and an infrastructure centre led by the University of Southampton in collaboration with Sheffield, Loughborough, Nottingham, and Heriot-Watt universities. These centres aim at developing technology and products for trains, systems, and infrastructure in order to deliver a better, more reliable, and more efficient railway.

At the digital systems centre in the University of Birmingham, various railway decarbonisation research projects have been carried out. For instance, in 2018, the SmartDrive package was developed using software created by Birmingham Centre for Railway Research and Education researchers. This package helped minimise energy usage within a fixed total journey time by instructing drivers to apply energy-efficient driving techniques. Its effectiveness has been verified by real daily operations in the Edinburgh Tram Line [21, 22]. In the CaFiBo project [23], the world's first carbon fibre bogie was developed by the University of Huddersfield's Institute for Railway Research in collaboration with a British company. The new bogie is lighter than conventional bogies and optimises vertical and transverse stiffness, which will, in turn, reduce energy consumption and hence global warming footprint. The University of Southampton also received funding from UK's Department for Transport to research and developed new masts made from advanced composite materials that have negative embedded carbon. The new masts would be used to replace carbon-intensive steel production [24], which could significantly reduce the cost of railway electrification.

In the Netherlands, several projects were carried out at Delft University of Technology continuing and extending the work of the EU project ON-TIME, in collaboration with ProRail and the Netherlands Railways. A pseudospectral optimal control approach was developed to compute optimal train trajectories for generic single-train and multi-train trajectory optimisation problems [25]. The train path envelope was further developed and included in energy-efficient train trajectory optimisation algorithms for C-DAS/ATO [26], as well as energy-efficient train timetables [27, 28]. In addition, in collaboration with the Dutch regional railway undertaking Arriva, a project was carried out to identify and evaluate solutions for replacing diesel traction on non-electrified railways by alternative catenary-free propulsion systems and low/zero emission energy carriers to improve energy efficiency and reducing greenhouse gas emissions. The OPEUS simulation model was applied in a Well-to-Wheel analysis of various alternative systems including alternative Energy Storage Systems and hydrogen-powered propulsion systems [29, 30].

To get on track with Net Zero Emissions by 2060, the Chinese government carried out a "14th Five-Year Plan" Railway Science and Technology Innovation Plan, which scheduled a series of projects to reduce the energy consumption of railway. These projects include: building an energy-efficient system for the operation and control of multiple types of trains, developing energy supply and management technologies that match the layout of railway facilities, and exploring innovative traction power technologies such as energy storage devices and fuel cells [31].

In 2019, Beijing Jiaotong University started a project, Investigation into Intelligent Control of Heavy Haul Trains on Long and Steep Downhill Section, which was funded by the National Natural Science Foundation of China. This project focused on the core issues in the driving control of heavy haul trains, including the longitudinal impulse transmission mechanism, the driving control modelling, the decision-making of the cycle braking, and intelligent control approaches. The main research covered modelling of the heavy haul train operation, analysis of longitudinal impulse transmission mechanism, optimisation modelling, and intelligent control approaches. Through this project, a theoretical foundation and support were gained for implementing intelligent control of heavy-duty trains and enhancing the smooth, safe, and effective operation of heavy-duty railways [32].

From 2017 to 2019, the project, Integrate Train Energy Efficiency Optimisation Model Basing on On-Board Energy Storage Devices was carried out by the South China University of Technology. In this project, a Mixed Integer Linear Programming (MILP) technique was used to optimise an integrated energy-efficiency optimisation model for trains with onboard ESDs. This research mainly addressed two issues: first, the optimisation of charging and discharging strategies for ESDs under the constraints of train operations; second, the optimisation of train operations considering the train's traction system characteristics. By considering both the constraints of train traction systems and onboard ESDs, this project further improved the energy efficiency of train operations. The outcomes of this project would significantly contribute to the development of energy efficiency improvement technologies in urban rail transit systems [33].

Australian rail research centres and industry bodies are actively involved in developing products and technologies to enhance all aspects of rail transport [34]. The Energymiser System, developed by the University of South Australia, is an in-cabin Driver Advisory System (DAS), which provides real-time driver advice and web-based reports. The Centre for Railway Engineering (CRE), based at CQUniversity in Queensland, is applying a number of engineering disciplines to rail research. They are developing an Intelligent Train Monitor (ITM), an in-cabin device that provides the train driver with information about forces in the train. The ITM's software platform is able to display a variety of information in many different forms. It will also enable comments from train drivers and operators to be quickly incorporated into the system, which in turn saves energy consumption [35, 36]. In 2022, the technology company ABB also started a three-year project, which aims at capturing braking energy and returning it to the 1500 V DC wayside energy storage system (ESS) for the acceleration of other trains [37].

In Spain, the Comillas Pontifical University carried out a project to optimise the Automatic Train Operation (ATO) speed profiles on the metro line of Madrid, which is equipped with the Communications-based train control system (CBTC). The objectives of this project are to minimise energy consumption and generate the Pareto optimal curve [38]. This project considered the uncertainty of the train mass as a fuzzy number model. NSGA-II-F algorithm was applied to design the optimal speed profiles. This project has been realised on a real interstation in Metro de Madrid, showing that significant energy savings can be obtained by re-designing ATO speed profiles while taking advantage of the CBTC features (7–8%).

The Seoul Metro of South Korea joined the green approach by working together with the high-tech company EMERSON. EMERSON developed the SCADA software for Seoul Metro, to improve its energy efficiency. The SCADA software includes an entire management system and the station monitoring system, which can collect and monitor energy data from hundreds of stations, substations, and depots [39]. Its effectiveness in monitoring and analysing energy consumption has been shown. Moreover, Seoul Metro achieved a 4% reduction in total energy consumption with this software [40].

In 2009, a railway project supported by the Japan Science and Technology (JST) and Japanese Ministry of Land, Infrastructure was started. This project provides a detailed analysis of voltage drop and energy-saving effects when using the superconducting cable [41]. The main study focused on the energy analysis of superconducting power transmission on commercial railway lines. The results of this research showed that the energy-saving rate increases as the length of superconducting cable extends. While with a short cable, the voltage drop specific to railways could be reduced. This project provides a good solution to the problem of saving railway power and energy-related problems with superconducting feeder systems.

Even though numerous projects have been carried out to reduce railway energy consumption, which can be seen in Table 1.1, the study of innovative strategies and technologies is still attracting researchers across the world. In 2016, a comprehensive analysis of the railway was made by the UIC [42]. This report described the most recent research about the potential reduction of energy consumption, according to the technologies developed nowadays. Additionally, it also analysed practices carried out by railway undertakings that encouraged energy efficiency. The results show that the energy consumption in existing railway systems could be reduced by 10–30% with the application of smart energy management, 15% by reduction of the losses in the traction chain, and 7–15% by the inclusion of reversible substations. Therefore, research about the railway's power supply, traction, and regenerative braking system is still a promising way to improve its energy efficiency.

1.3 Energy Saving Methods in Railways

The energy consumption of a railway system is composed of many parts, which can be divided into traction energy consumption and non-traction energy consumption (see Fig. 1.6). The traction energy consumption represents the traction energy offered by substations for train operation, which includes the energy consumed by the onboard auxiliary systems, the traction energy used to overcome the motion resistance of rolling stock, the energy losses in the traction chain, etc. The non-traction energy consumption embraces all the energy utilised by different services ensuring the proper operation of urban rail systems. These typically comprise passenger stations, depots, and other infrastructure-related facilities such as signalling systems, tunnel ventilation fans, groundwater pumps, and tunnel lighting. Among the total energy consumption, traction energy consumption accounts for about 80% of the total energy consumption of the railway system, which is the focus of energy conservation research.

To reduce the traction and non-traction energy, Table 1.2 presents the main efficient measures proposed and implemented so far in the urban rail transit system, which can be divided into five categories: using regenerative braking, implementing eco-driving strategies, minimising traction losses, reducing the energy demand of comfort functions, and measuring and managing the energy flows efficiently. The first measure aims to recovering and reusing the vehicles' braking energy in the form of electricity. The synchronisation of train timetables, the usage of Energy Storage System (ESS), and the construction of reversible substations belong to this measure. Energy-efficient driving is the second energy-saving measure which refers to the group of techniques intended to operate rail vehicles as efficiently as possible while ensuring the safety and punctuality of services. Apart from energy consumption reduction, eco-driving strategies may also improve passenger comfort through smoother driving and reduce the wear of rolling components. Optimising the speed profiles, coasting and using the track gradients are the basic practices in eco-driving, supported by optimisation of timetables and railway traffic management. The energy-efficient traction systems can reduce the energy losses by optimising the parameters of the traction system. The fourth measure is reducing the energy demand of comfort functions, such as the optimal control of the fresh air supply. Measuring and managing the energy flows is the final measure including energy metering, local renewable power generation, and smart power management.

Table 1.1 Railway projects for energy-efficiency improvement

Project	Country/ region	Main contribution to improving energy efficiency	Duration
ON-TIME [10–12]	EU	Developed methods, system architectures and exchange data formats for train timetabling, traffic management and connected Driver Advisory Systems, integrating energy-efficient train trajectories throughout	2011–2014
OPEUS [15]	EU	1. Developed simulation methods and associated tools to evaluate and optimise energy consumption 2. Provided robust, stable, and readily available assessment methodology	2016–2019
In2Stempo [16]	EU	1. Developed a smart railway power grid 2. Constructed a fine mapping of energy flows	2017–2022
Railenergy [17]	EU	1. Addressed Energy Efficiency Issues for Integrated Rail Systems 2. Developed new verification standards for energy performance of railway products and services	2006–2010
MERLIN [18]	EU	Investigated and demonstrated the viability of an integrated management system to optimise energy usage in electric mainline railway systems	2012–2015
OSIRIS [19]	EU	1. Identified operational and technical innovations that reduce energy costs in running rail systems 2. Developed innovative methodology for simulating, evaluating, and optimising energy consumption in urban rail systems	2012–2015
SmartDrive [21, 22]	UK	Developed the SmartDrive package to achieve the application of an energy-efficient driving strategy	2019
CaFiBo [23]	UK	Developed the carbon fibre rail bogie, which can be used to reduce the mass of the rolling stock	2017–2019
Innovative composite mast for greener electrification [24]	UK	Developed new masts made from advanced composite materials to reduce the cost of electrification	2021
“14th five-year plan” railway science and technology innovation plan [31]	CN	Carry out a series of research plans to reduce the energy consumption of railway	2021–2026

(continued)

Table 1.1 (continued)

Project	Country/ region	Main contribution to improving energy efficiency	Duration
Investigation into intelligent control of heavy haul trains on long and steep downhill section [32]	CN	Obtained a theoretical basis and support for realizing intelligent control and improving the safe, smooth, and efficient operation of heavy-haul railway	2019–2021
Integrate train energy efficiency optimisation model basing on on-board energy storage devices [33]	CN	Developed an integrated energy-efficiency optimisation model for trains with onboard ESDs, and obtained the optimisation results with mixed integer linear programming (MILP)	2017–2019
ITM [35, 36]	AUS	Developed an in-cab system that can help with energy-efficient driving	1994–2008
Captures, stores and regenerates braking energy [37]	AUS	Aims to capture and store regenerative braking energy in the wayside energy storage system	2022–2025
Optimal design of energy-efficient ATO CBTC driving for metro lines [38]	ESP	1. Re-designed the ATO speed profiles and achieved energy saving 2. Provided a well-distributed pseudo-Pareto front	2014
Seoul metro and Emerson's cooperative project [39, 40]	KOR	1. Monitor and analysis of energy consumption 2. Reduction of the total energy consumption	2019
Energy analysis of superconducting power transmission installed on the commercial railway line [41]	JP	Achieved energy saving and suppression of voltage drop on city rail line model	2020

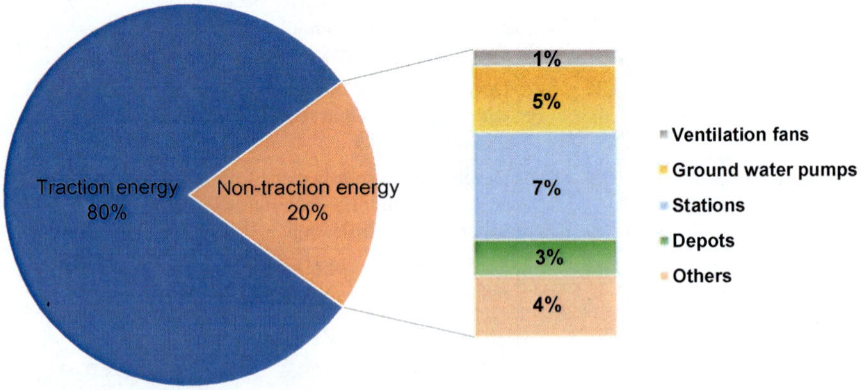

Fig. 1.6 Distribution of non-traction energy in railway systems [43]

Table 1.2 Main actions to save energy in urban rail [43]

Measures	Applied range	Type of measures	Objectives of measures
Timetable optimisation	Whole system	Operational measure	Regenerative braking
Renewable energy generation	Whole system	Technological measure	Measurement and management
Smart energy management	Whole system	Operational and technological measure	Measurement and management
Optimised traffic management	Whole system	Operational and technological measure	Energy-efficient driving
Energy metering	Whole system	Technological measure	Measurement and management
Passenger management in stations	Infrastructure	Operational and technological measure	Passenger management in stations
Reversable substations	Infrastructure	Technological measure	Regenerative braking
Reduced power supply losses	Infrastructure	Operational and technological measure	Traction efficiency
Low-energy tunnel cooling	Infrastructure	Technological measure	Passenger management in stations
Lighting and HVAC in stations	Infrastructure	Operational and technological measure	Passenger management in stations
Wayside ESS	Infrastructure	Technological measure	Regenerative braking
Optimised traction software	Rolling stock	Operational measure	Traction efficiency
Timetable optimisation	Rolling stock	Operational and technological measure	Passenger management in stations
Mass reduction	Rolling stock	Technological measure	Traction efficiency
ATO	Rolling stock	Technological measure	Energy-efficient driving
Thermal insulation	Rolling stock	Technological measure	Passenger management in stations
Eco-driving techniques	Rolling stock	Operational measure	Energy-efficient driving
PMSM	Rolling stock	Technological measure	Traction efficiency
On-board ESS	Rolling stock	Technological measure	Regenerative braking
Lighting and HVAC in Service	Rolling stock	Operational and technological measure	Passenger management in stations
DAS	Rolling stock	Technological measure	Energy-efficient driving

The relationship among the first four measures is shown in Fig. 1.7, while the general evaluation of each energy efficiency measure is shown in Table 1.3. According to the energy saving potential index, most methods of the regenerative braking measure and energy-efficient driving measure have an energy-saving potential of more than 5%, while only two methods of traction efficiency and comfort function measure may achieve an energy saving potential of more than 5%. Therefore, the optimisation measures on train operation need to be focused on to increase

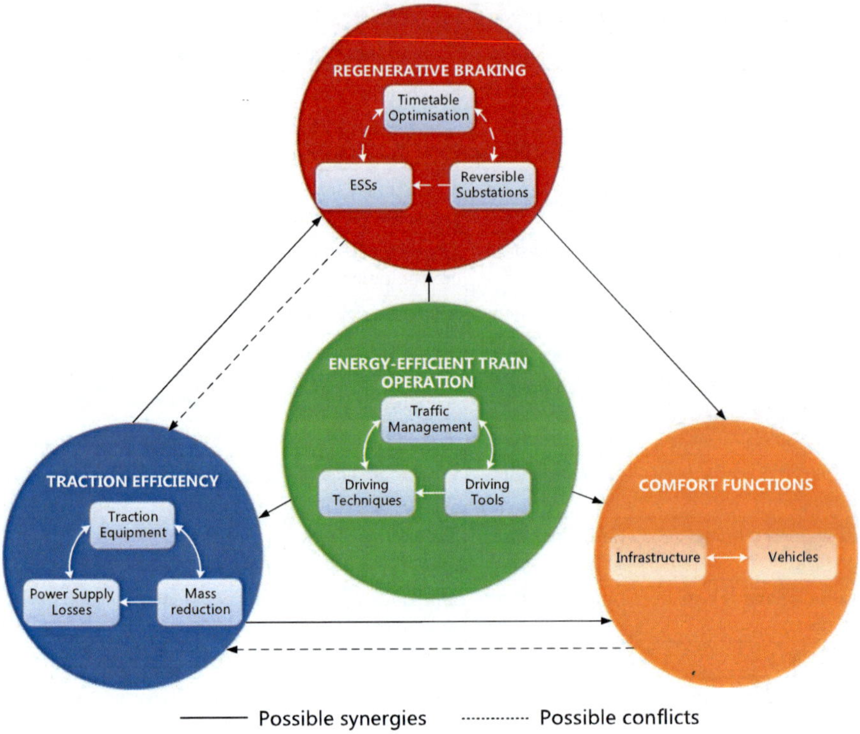

Fig. 1.7 Interdependences between energy efficiency measures [43]

the energy-saving effect by improving the utilization rate of regenerative energy and reducing the traction energy consumption.

1.4 Book Chapter Structure

Due to the significance of rail energy and the high potential to reduce the energy consumption in railway systems, this book further illustrates the energy-efficient train operation solutions for railway systems. This book is closely related to the energy conservation problem of the rail transit system by focusing on reducing the energy consumption for train operation. In this book, the whole systems processes of train operation with optimisation solutions are analysed and illustrated. The remainder of this book is structured as follows.

Chapter 2 introduces the relationship between train operation and energy consumption under different energy-efficient strategies. This relationship will serve as the basis for the energy-efficient optimisation methods in the subsequent chapters. Specifically, in the train-based energy-saving strategy that aims to minimise the net

Table 1.3 General evaluation of energy efficiency measures in urban rail systems [43]

Measures			Energy saving potential (%)	Suitability for existing systems	Investment cost
Cluster	Category	Solution			
Regenerative braking	Timetable optimisation		1–10	High	Low
	ESS	On-board	5–25	Medium	High
		Stationary	5–25	High	High
	Reversible substations		5–20	High	High
Energy-efficient driving	Eco-driving techniques	Coasting, optimised speed profile, use of track gradients	5–10	High	Low
	Eco-driving tools	DAS	5–15	High	Medium
		ATO	5–15	Medium	High
Traction efficiency	Power supply network	Higher line voltage	1–5	Low	High
		Lower resistance conductors	1–5	Low	High
	Traction equipment	PMSM	5–10	High	High
		Software optimisation	1–5	High	Low
	Mass reduction	Materials substitution	1–10	High	Medium
Comfort functions	Vehicles	Thermal insulation	1–5	High	Medium
		Heap pump	1–5	Medium	Medium
		LEDs	1–5	High	Medium
		HVAC and lighting control in service	1–5	High	Low
		HVAC and lighting control in parked mode	1–5	High	Low
	Infrastructure	Low-energy tunnel cooling	1–5	Low	High
		Geothermal heat pumps	1–5	Medium	Medium
		Control of HVAC, lighting and passenger conveyor systems	1–5	High	Low
		LEDs	1–5	High	Medium

energy consumption, four methods are introduced, including energy-efficient train control, energy-efficient train timetabling, integrated optimisation for regenerative braking, and energy-efficient driving considering energy storage systems. Then, the substation-based energy-saving strategy whose goal is to reduce the electric energy offered by substations, is briefly presented.

Chapter 3 formulates an optimisation model for energy-efficient train operation. The motion of the train is described by a pair of differential equations, with parameters including the tractive and braking effort curves, resistance forces, track gradient forces, and track curvature forces. Speed limits and timing requirements are imposed as constraints on the motion. By solving the optimisation model with the Pontryagin's maximum principle, five optimal driving modes including power, hold, coast, regen, and brake are achieved. Furthermore, these necessary conditions for optimal control are used to determine the optimal sequence of control modes for a journey, and when the control should be switched between modes.

Chapter 4 applies train trajectory optimisation to compute the minimum-time and energy-efficient train trajectories between two and more stops, as well as the optimal running time supplements in timetables over corridors with multiple stops. Also the trade-off is considered between energy-efficient train trajectories and the infrastructure occupation of multiple trains on heterogeneous train traffic corridors. The results are compared to other train-driving strategies used in practice. Realistic conditions are considered, including varying gradients and speed restrictions, and heterogeneous mainline railway traffic.

Chapter 5 focuses on the integrated driving strategy and train timetable synchronization method. Firstly, the mathematical models are formulated to calculate the amount of traction energy and reused regenerative braking energy (RBE). Then, two integrated timetabling approaches are introduced. In the first approach, the arrival time and departure time of trains are synchronized with consideration of the driving strategy to efficiently reuse the produced RBE. The second approach matches traction/braking regimes during the inter-station train operation, to get a better energy-saving effect. Both integrated timetabling approaches are verified by metro line examples based on real-world data.

Chapter 6 discusses the main technologies, the modelling, and control methods of energy storage systems. A case study is presented where different scenarios of energy storage and receptivity to regenerated energy are analysed based on the characteristics of a real line of the Madrid Underground. The influence of energy storage devices in energy consumption reduction and the optimal design of ATO speed profiles is evaluated.

Chapter 7 presents the development of energy evaluation simulation of electric railway systems. The train movement model and railway power network model are integrated into the simulator. Based on the power network model, this chapter also analyses the energy consumption of railway systems with regenerating trains, including the energy supplied by substations, used in power transmission networks, consumed by monitoring trains, and regenerated by braking trains. Finally, a case study of the Beijing Yizhuang Subway Line is conducted, which indicates that the available regenerative braking energy and total substation energy consumption vary with timetables.

Chapter 8 gives the basic conclusions about energy-efficient train operation covering energy-efficient train driving, energy-efficient train timetabling, regenerative braking, energy storage systems and power supply networks. This chapter also provides recommendations for further research, which includes the interaction of connected driver advisory systems (C-DAS) and automatic train operation (ATO) with railway traffic management systems, cooperative train control in platoons of virtually coupled trains, digital twin technology and particularly its application to power supply systems, and the interaction between the railway network with the electrical power grid and renewable energy generation.

References

1. IEA (International Energy Agency) and UIC (International Union of Railways) (2019) The future of rail opportunities for energy and the environment
2. IEA (International Energy Agency) and UIC (International Union of Railways) (2017) Railway handbook on energy consumption and CO_2 emissions
3. Fouracre P, Dunkerley C, Gardner G (2003) Mass rapid transit systems for cities in the developing world. Transp Rev 23(3):299–310
4. Comparisons of cities' transportation modal shares and post-coronavirus prospects. https://www.sc-abeam.com/and_mobility/en/article/20201203-01/
5. Ten charts to help you understand the market environment and railroad transportation industry development possibilities in China in 2021. https://www.qianzhan.com/analyst/detail/220/210604-609e6443.html
6. Department of Infrastructure and Regional Development (2014) Australian freight transport overview
7. UIC, and CER (2012) Moving towards sustainable mobility: a strategy for 2030 and beyond for the European railway sector. UIC Communications Department
8. Goverde RMP, Bešinović N, Binder A, Cacchiani V, Quaglietta E, Roberti R, Toth P (2016) A three-level framework for performance-based railway timetabling. Transp Res Part C Emerg Technol 67:62–83
9. Quaglietta E, Pellegrini P, Goverde RMP, Albrecht T, Jaekel B, Marlière G, Rodriguez J, Dollevoet T, Ambrogio B, Carcasole D, Giaroli M, Nicholson G (2016) The ON-TIME real-time railway traffic management framework: a proof-of-concept using a scalable standardised data communication architecture. Transp Res Part C Emerg Technol 63:23–50
10. Panou K, Tzieropoulos P, Emery D (2013) Railway driver advice systems: evaluation of methods, tools and systems. J Rail Transp Plann Manage 3(4):150–162
11. ON-TIME. https://cordis.europa.eu/project/id/285243
12. ON-TIME (2014) Specification of a driving advisory systems (DAS) data format, p. Deliverable 6.1
13. UIC. Data exchange with driver advisory systems (DAS) following the SFERA protocol
14. Wang Z, Quaglietta E, Bartholomeus MGP, Goverde RMP (2022) Assessment of architectures for automatic train operation driving functions. J Rail Transp Plann Manage 24:100352
15. OPEUS Project. https://opeus-project.eu/
16. In2Stempo Project. https://projects.shift2rail.org/s2r_ip3_n.aspx?p=IN2stempo
17. Railenergy Project. http://www.railenergy.org/
18. MERLIN Project. http://www.merlin-rail.eu/
19. OSIRIS Project. http://www.osirisrail.eu

20. RSSB launches £92m British rail research network. https://www.railjournal.com/regions/europe/rssb-launches-92m-british-rail-research-network/
21. Birmingham centre for railway research and education industry collaborator commended at the global light rail awards. https://www.birmingham.ac.uk/news/2018/birmingham-centre-for-railway-research-and-education-industry-collaborator-commended-at-the-global-light-rail-awards
22. Tian Z, Zhao N, Hillmansen S, Roberts C, Dowens T, Kerr C (2019) SmartDrive: traction energy optimization and applications in rail systems. IEEE Trans Intell Transp Syst 20(7):2764–2773
23. World's first carbon fibre rail bogie unveiled on campus. https://www.hud.ac.uk/news/2019/december/carbon-fibre-rail-bogie-unveiled/
24. University of Southampton wins funding for research to reduce the environmental impacts of rail travel. https://www.southampton.ac.uk/news/2021/07/greener-railway-masts.page
25. Goverde RMP, Scheepmaker GM, Wang P (2021) Pseudospectral optimal train control. Euro J Oper Res 292(1):353–375
26. Wang P, Goverde RMP (2016) Multiple-phase train trajectory optimization with signalling and operational constraints. Transp Res Part C Emerg Technol
27. Scheepmaker GM, Goverde RM (2021) Multi-objective railway timetabling including energy-efficient train trajectory optimization. EJTIR 21(4):1–42
28. Wang P, Goverde RMP (2019) Multi-train trajectory optimization for energy-efficient timetabling. Euro J Oper Res 272(2):621–635
29. Kapetanović M, Núñez A, van Oort N, Goverde RMP (2021) Reducing fuel consumption and related emissions through optimal sizing of energy storage systems for diesel-electric trains. Appl Energy 294:117018
30. Kapetanović M, Núñez A, van Oort N, Goverde RMP (2022) Analysis of hydrogen-powered propulsion system alternatives for diesel-electric regional trains. J Rail Transp Plann Manage 23:100338
31. "14th five-year plan" railway science and technology innovation plan. http://www.gov.cn/zhengce/zhengceku/2021-12/24/content_5664357.htm
32. Investigation into intelligent control of heavy haul trains on long and steep downhill section. https://kd.nsfc.gov.cn/finalDetails?id=f2b1e1f35449254e089c58d3ecb4ce00
33. Integrate train energy efficiency optimisation model basing on on-board energy storage devices. https://kd.nsfc.gov.cn/finalDetails?id=9473c440b42a81e0e963d25540e654a9
34. Australian Trade Commission and Australia UNLIMITED (2013) Heavy haul international and freight rail
35. Gorman M, Roach D, Cole C, McLeod T. The development of a train dynamics monitor
36. Mcclanachan M, Phelan J. Intelligent train monitor developments
37. Australia's ABB captures, stores and regenerates braking energy. https://www.railway-technology.com/analysis/australias-abb-captures-stores-and-regenerates-braking-energy/
38. Carvajal-Carreño W, Cucala AP, Fernández-Cardador A (2014) Optimal design of energy-efficient ATO CBTC driving for metro lines based on NSGA-II with fuzzy parameters. Eng Appl Artif Intell 36:164–177
39. Inter Electric (2019) Seoul metro joins the green approach
40. EMERSON (2021) Seoul Metro uses Emerson SCADA software to monitor and reduce electricity consumption across rail network
41. Tomita M, Fukumoto Y, Ishihara A, Suzuki K, Akasaka T, Kobayashi Y, Onji T, Arai Y (2020) Energy analysis of superconducting power transmission installed on the commercial railway line. Energy 209:118318
42. UIC (International Union of Railways) (2016) Technologies and potential developments for energy efficiency and CO_2 reductions in rail systems
43. González-Gil A, Palacin R, Batty P, Powell JP (2014) A systems approach to reduce urban rail energy consumption. Energy Convers Manage 80:509–524

Chapter 2
Energy-Efficient Strategies for Train Operation

Shuai Su, Rob M. P. Goverde, Xuekai Wang, and Zhongbei Tian

2.1 Introduction

The energy consumption of train operation accounts for a large proportion of the total energy consumption in the railways, which needs to be reduced to further improve environmental sustainability and save operating cost. In this chapter, the relationship between the energy consumption and train operation will be introduced in detail, as the theoretical basis for various approaches to improve energy-efficient train operation.

2.1.1 Background of Train Operation

The basis of a railway transportation system is formed by an extensive planning stage, resulting in a timetable, rolling stock plans, and crew duties [6]. The railway train timetable is designed considering one or more of the objectives including energy consumption [15], train travel time [41], etc. Based on the train timetable, the rolling

S. Su (✉) · X. Wang
Beijing Jiaotong University, Beijing, China
e-mail: shuaisu@bjtu.edu.cn

X. Wang
e-mail: 19111067@bjtu.edu.cn

R. M. P. Goverde
Delft University of Technology, Delft, Netherlands
e-mail: R.M.P.Goverde@tudelft.nl

Z. Tian
University of Birmingham, Birmingham, UK
e-mail: z.tian@bham.ac.uk

S. Su et al., *Energy-Efficient Train Operation*, Lecture Notes in Mobility,
https://doi.org/10.1007/978-3-031-34656-9_2

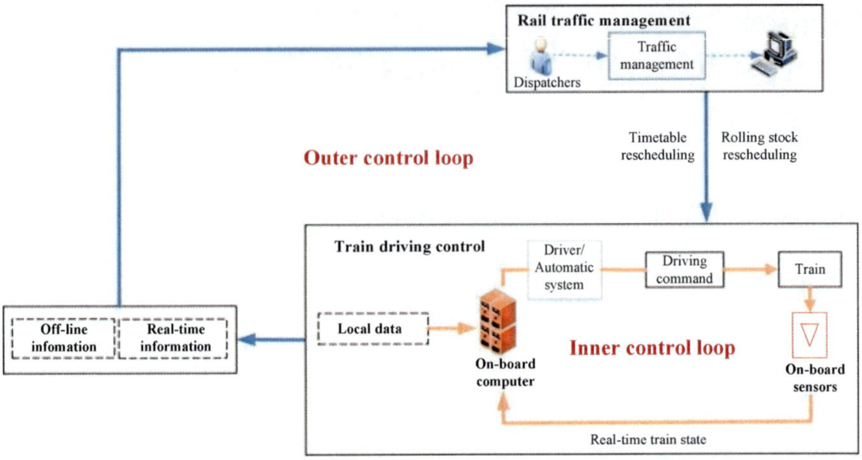

Fig. 2.1 Relation between railway traffic management and train operation [39]

stock plans and crew duties are assigned by railway undertakings [5]. The planning phase is carried out before the real-time train operation.

In the real-time railway operation, the occurrences of unexpected disturbances and disruptions may result in the infeasibilities of the original plans [6]. To ensure normal train operation, two control tasks are required [39]. The "rail traffic management" needs to be taken to reschedule or adjust the train timetable and rolling stock plan with real-time train information and estimations of disturbances/disruptions. In this way, the impacts of disturbances/disruptions are reduced by dispatchers in a railway traffic control centre. The rail traffic management can be illustrated by an outer control as shown in Fig. 2.1. To execute the scheduled (or rescheduled) plans, another basic task in the real-time railway operation is "train driving control", which focuses on the safe and efficient train movements on a microscopic level by determining the train control commands, i.e., applied traction and braking forces [39]. As shown in Fig. 2.1, train driving control is executed by an inner control loop based on the scheduled (or rescheduled) plans and real-time train information for each single train.

2.1.2 *Approaches of Energy-Efficient Train Operation*

As shown in Fig. 2.2, energy-efficient train operation can be divided into two aspects according to different energy-saving objectives. The first aspect is the train-based energy-efficient strategy. In this aspect, the objective is to reduce the total net energy consumption of train operation. In other words, only the mechanical energy required by the train operation is considered. The factors directly related to the train operation (e.g., running resistance in Fig. 2.3) need to be taken into account. The other aspect is the substation-based energy-efficient strategy. In this aspect, increasing the efficiency

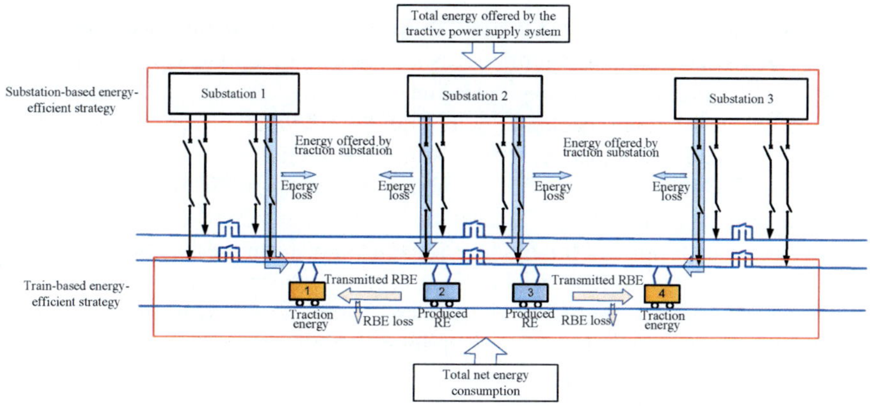

Fig. 2.2 Two aspects of energy-efficient train operation based on different objectives

Fig. 2.3 Diagram of traction energy consumption distribution

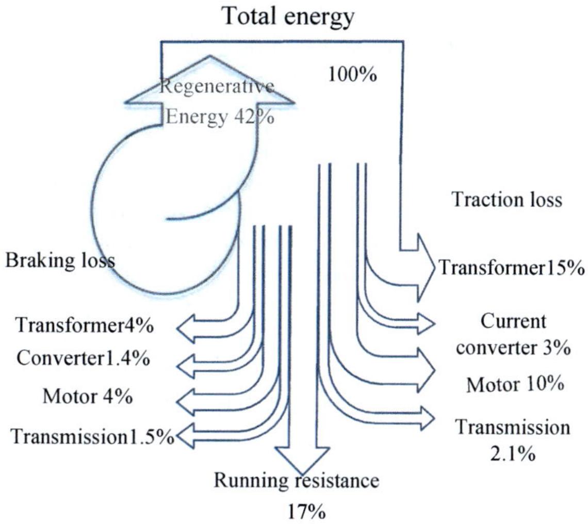

of the power supply system minimising the total energy offered by the Traction Power Supply System (TPSS) is the objective. Therefore, not only the energy required by the operated trains but also the energy flow in the TPSS need to be considered simultaneously. As shown in Fig. 2.3, the electric energy obtained by a train from the TPSS is lost in transformers, converters, motors, and transmissions (about 30%) during the process of train traction [13]. In the substation-based strategy, it is required to reduce such loss by energy-efficient optimisation.

The train-based energy-efficient strategy aims to reduce the total net energy consumption of all the operated trains, which is the difference between the traction energy and the reused Regenerative Braking Energy (RBE). For the traction energy, some is used to overcome the resistance in the traction process, and the rest is converted into the kinetic energy to accelerate the train. During the cruising regime, traction energy is employed to overcome the running resistance while the kinetic energy remains constant. If friction braking is applied in the braking regime, the kinetic energy will be directly converted into heat energy through the friction between the rail and the brake shoe, which means that about 50% of the total traction energy consumed before this braking regime will be lost. Most modern urban rail vehicles have the ability of "regenerative braking" [12]. This electric braking method makes the motor act as a generator when applying the braking force, to transform the kinetic energy into electricity. The produced RBE can be stored in energy storage systems (ESSs) for the use by other devices (including trains), or can be transmitted to other trains through the catenary system. By using RBE, the train can still use the conventional air braking method to provide fast and powerful braking force. Especially, when the train is running at a low speed, the RBE will convert most of the kinetic energy into electric energy, which can account for 80% of the kinetic energy. This amount of RBE is also equivalent to about 40% of the traction energy obtained from the TPSS. Therefore, from the perspective of the energy flow in the rail system, there are two main ways to reduce the energy consumption of train operation: (1) reduce the required traction energy during the train operation; (2) increase the amount of reused RBE.

For the railway system, the train timetable and the inter-station driving strategy jointly determine the train operation process. Among them, the train timetable designed in the planning stage describes the process of the train operation, including the order of the operated trains, the arrival and departure time at each station, the inter-station running time, the dwell time at the station, etc. In the real-time train driving control, the train is manually or automatically controlled in each segment according to the driving strategy, which depicts the recommended train speed and force of each time instant. Both the train timetable and the driving strategy determine the amount the traction energy and the reused RBE.

The relationship among different energy-efficient strategies can be summarized as Fig. 2.4. The train-based energy-efficient strategy can be divided into two types: only reducing traction energy consumption, and simultaneously reducing traction energy consumption and increasing reused RBE. In the first type, the driving strategy is directly related to the required traction energy consumption of a single train. Therefore, energy-efficient train driving control is an efficient way to reduce the energy consumption of a single train in one segment. The above method will be introduced in Chap. 3. Based on the efficient driving strategy, the train timetable can be optimised by optimally allocating running time supplement and reducing the route occupation conflicts, to minimise the traction energy consumption of multiple trains in a rail network. The related train timetable optimisation methods will be introduced in Chap. 4.

The second type of train-based energy-efficient strategy is to efficiently reuse the produced RBE. According to the method to reuse the produced RBE, the energy-efficient strategy can be further divided into two kinds. If the produced RBE feeds back to the catenary system to be used by the nearby traction trains, then the key to efficient reuse RBE is to match the times for adjacent trains to apply the traction/braking regimes. In this way, the reuse of RBE is jointly related to the train

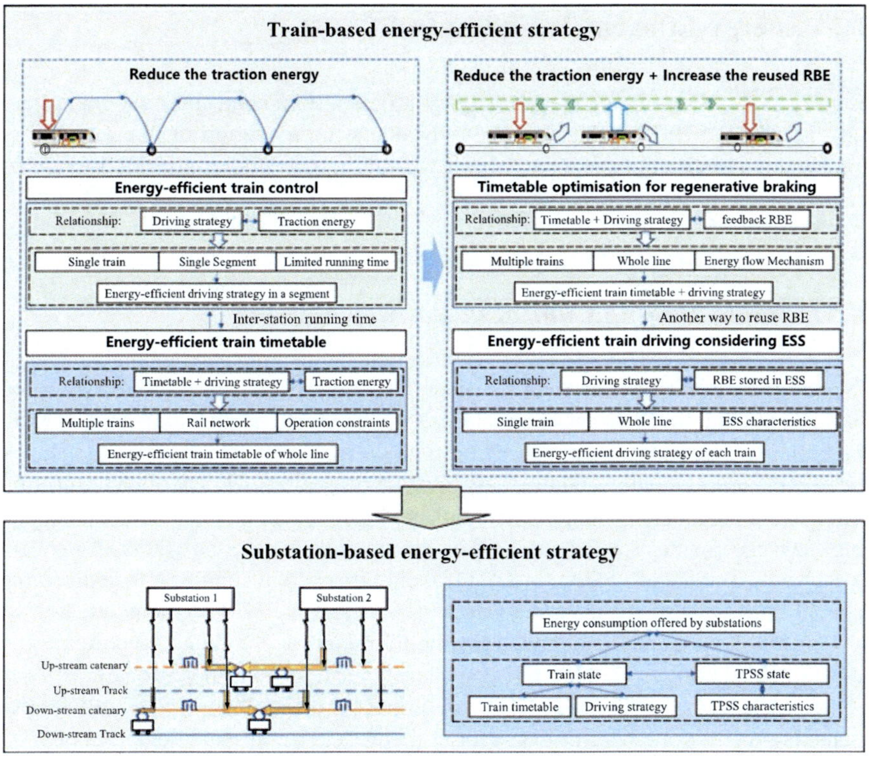

Fig. 2.4 Relationship between train operation and energy consumption

timetable and the driving strategy. As a result, an integrated optimisation method is necessary. This method will be introduced in Chap. 5. Besides, the produced RBE can also be stored in an ESS for the reuse at another time. Based on the characteristics of ESS, the driving strategy of a single train can be optimised to determine when to charge or discharge the ESS. This energy-efficient method will be detailed in Chap. 6.

Compared with the train-based energy-efficient strategy, the electric energy offered by the substations can be precisely calculated in the substation-based energy-efficient strategy by considering both the train state and the TPSS state. The train state at each instant is determined by the train timetable and driving strategy of each train, while the TPSS state is influenced by the train state and TPSS parameters. As a result, the substation-based approach can be achieved by jointly optimising the train timetable and driving strategy. The relevant content will be introduced in Chap. 7.

Next, the principle of each energy-efficient train operation strategy will be briefly described in Sects. 2.2–2.6.

2.2 Energy-Efficient Train Control

In this section, the principles of the energy-efficient train control for an single train will be introduced. The objective only considers the reduction of traction energy consumption, and will be used as a basis for subsequent studies considering multi-train or multi-station conditions.

2.2.1 Train Driving Control in Railway Systems

To achieve safe and efficient train movement, a train can be controlled and operated by drivers manually or by automatic driving control systems. In the manual driving, the train is operated by drivers based on track-side interlocking and blocking combined with train signalling and supervisory devices [10, 24]. The manual driving relies heavily on driver's experience and is usually vulnerable to external factors (e.g., driver's mental state, bad weather that may influence the sights of drivers) [39]. Automatic Train Operation (ATO) is an emerging technology to replace the manual driving in many railway systems [10, 24, 25, 37]. The main function of ATO is to automatically generate real-time train control commands according to a recommended train trajectory while considering the timetable and the characteristics of the train and infrastructure [36]. At present, ATO technologies are mainly implemented in metro systems. Besides, a lot of theoretical studies and field tests of ATO technologies are ongoing for other railway systems [39].

In metro lines, the ATO function is usually achieved by an Automatic Train Control (ATC) system, which is an integrated signalling system that combines railway train control, supervision and management to guarantee the safe and efficient movement of trains. As illustrated in Fig. 2.5, ATC contains three subsystems, i.e., Automatic Train Supervision (ATS), Automatic Train Protection (ATP), and ATO [34]. The ATO subsystem can use computer programming and control techniques to automatically control the train movements with the supervision of the ATS and ATP subsystems. Based on the monitored train state and schedule, ATS gives train routing and schedule adherence instructions to the ATO subsystem for deciding the train control commands. Meanwhile, the ATP guarantees the safe train separation and overspeed protection [39].

As for mainline railway systems, a set of ATO-over-ETCS system requirement specifications has been developed in Europe [42]. The train driving functions of ATO-over-ETCS is shown in Fig. 2.6. In this architecture, Traffic Management System (TMS) offers the infrastructure and timetable information including the route, the targets at timing points (e.g., arrival time, departure time), temporary speed restrictions, and low adhesion (if applicable) to the ATO-TrackSide (ATO-TS) [36]. The above information is transformed and sent to the ATO-OnBoard (ATO-OB). Then, the ATO-OB can execute four parts of driving function including timetable

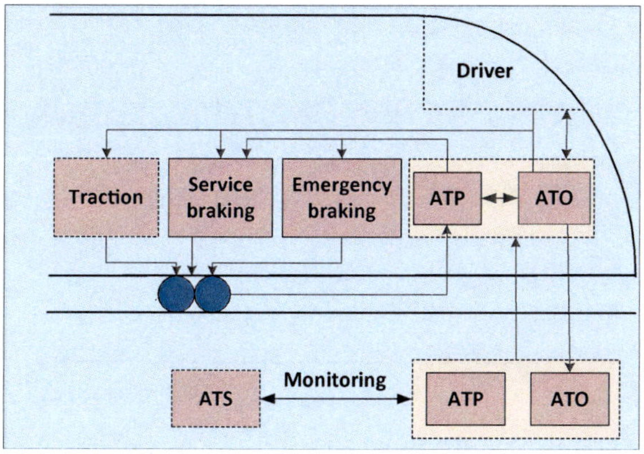

Fig. 2.5 Illustration of ATC structure in metro systems [39]

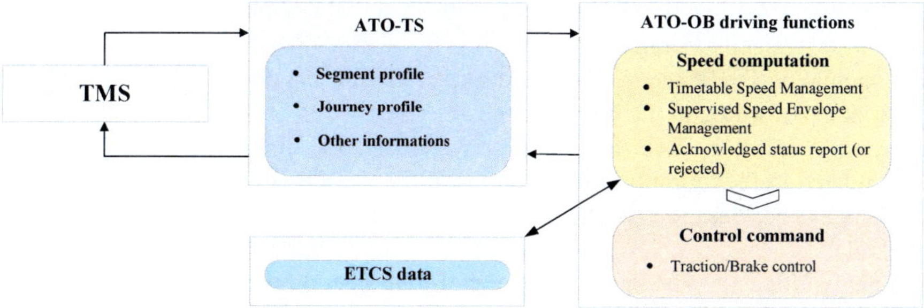

Fig. 2.6 Driving functions of ATO-over-ETCS [36]

speed management, supervised speed envelope management, automatic train stopping management, and ATO traction/brake control [11, 36].

To implement the automatic train driving function, the calculation and tracking of a recommended speed profile are two key functions of ATO. Before the train departs from a station, a recommended speed profile for the train operation (see red curve in Fig. 2.7) is calculated on the basis of segment and journey profile information, e.g., distance to the next station, line gradients, speed limits, etc. During the train operation, the ATO subsystem receives real-time data (e.g., train position and speed) and line information (e.g., speed limits, gradients) by on-board sensors and radio communication with the trackside [39]. The monitored train speed is compared with the recommended speed to determine the control command (i.e., applied traction and braking forces). In this way, the train can track the recommended speed profile as close as possible.

Performance indexes of the recommended speed profile generated by an ATO subsystem generally include stopping accuracy, punctuality, comfort level, and energy efficiency [39].

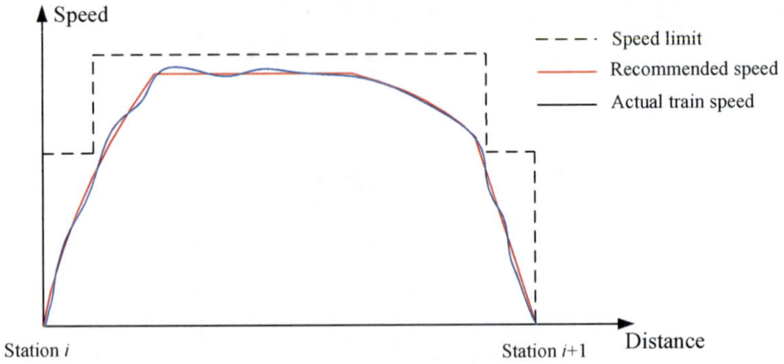

Fig. 2.7 Train speed control by ATO on a segment

- Stopping accuracy: the distance between the actual stopping position of the train and the planned stopping position.
- Punctuality: guarantee that the train arrives at and (or) departs from each station on time according to the train timetable.
- Comfort level: as indicated in railway standard [4], the comfort of passengers is related to the vibrations and motions of the vehicle. For example, the average change of acceleration and deceleration can be an indicator to measure the comfort level.
- Energy efficiency: the driving strategy of train operation is related to the traction energy consumption and reused RBE. The detailed relationship will be introduced in this chapter later.

2.2.2 Relationship Between Driving Strategy and Energy Consumption

As introduced above, the train is operated according to a driving strategy in each segment. Usually, the train need to complete the operation between stations based on a scheduled running time. Even if the inter-station running time is fixed, the train also has many possible driving strategies in a segment. Although each driving strategy can ensure that the train arrives at the next station on time, the corresponding traction energy consumptions are normally different. For example, Fig. 2.8 shows three driving strategies. The inter-station running time for them are the same, while the traction energy consumption is different, i.e., driving strategy 1 corresponds the lowest traction energy among them. There are three main factors that affect the energy consumption of the driving strategy: the category of driving regimes, sequence of driving regimes, and switching point of driving regimes.

Firstly, the driving regime for the train operation can be divided into 4 kinds according to the traction level and braking level.

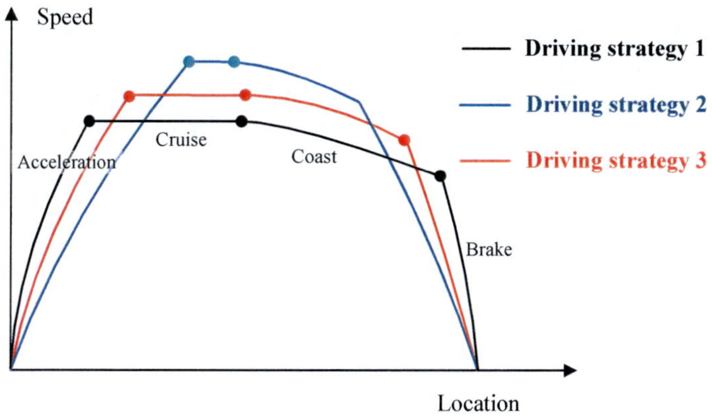

Fig. 2.8 Three recommended speed profiles

- Acceleration regime: Under the acceleration regime, the traction force is applied by the train, which can be controlled by adjusting the traction level. The traction level is the ratio between the actual output traction force and the maximum traction force. In this regime, the train will get electrical energy from the TPSS in order to supply the traction force needed to overcome various resistances. Therefore, traction energy will be consumed in this regime.
- Cruise regime: during the cruise regime, the train runs with a constant speed. In this way, the energy consumption of the train is closely related to the running resistance. When the total resistance is positive, the direction of resistance is opposite to that of the train operation. Therefore, a part of the traction force is required to keep a constant speed. When the running resistance is negative, it is in the same direction as the train operation. The braking force will be applied and no electric energy will be consumed.
- Coast regime: during the coast regime, the traction level and braking level are both zero. Then, the resultant force on the train is the same as the total resistance. The train will slow down if the total resistance is positive. On the contrary, the train accelerates when the resistance is negative. Because there is no control force exerted on the train, there is no trcation energy consumption during the train operation.
- Braking regime: the braking force is applied during the braking regime. Therefore, there is no traction energy consumption. Besides, the kinetic energy of the train can be converted into electric energy if regenerative braking is applied. The RBE can be stored in an ESS or fed back to the catenary for other trains or auxiliary equipments to use.

Apart from the category of driving regimes, the actual sequence and the switching point locations affect the traction energy consumption. Normally, the train will take the acceleration regime after departing from a station to accelerate. The braking regime is applied when decelerating to arrive at next station. In between, there are a variety of regime sequences to choose from, e.g., traction-coast-traction-coast-... A different sequence always implies different traction energy consumption. Moreover, the location of the switching points determines the initial and final train speed in each driving regime, which is also related to the amount of traction energy. As a result, the parameters of the driving regimes need to be optimised in the energy-efficient train control. Next, the factors to be considered in the energy-efficient train control will be introduced.

2.2.3 *Factors Related to the Train Motion*

Three main aspects of factors need to be considered during the optimisation of train energy-efficient driving.

- Vehicle characteristics: due to the impacts of vehicle parameters, the maximum traction force and the maximum braking force applied on the train are limited. Besides, when the train speed increases, the maximum traction/braking force will generally decrease.
- Track characteristics: affected by the undulations of the track, the train will be affected by gradient forces during the operation. The track curvature force also exists between the wheel and the track side when the train runs on the curve. Generally, the gradient force and curvature force are not related to the train speed.
- Basic resistance characteristics: the basic resistance on the train is mainly due to the internal or external friction and impact of the train. The main components of resistance include the bearing resistance, the sliding and rolling friction resistances between wheels and track, impact vibration resistance, and air resistance. The basic resistance changes with the speed of the train. At low speed, the bearing resistance accounts for a large proportion. With the increase of speed, the proportion of other resistances gradually increases. Because the relationship between these forces is very complex, it is difficult to use a theoretical formula to solve in practice, so an empirical formula is usually used to calculate the basic resistance.

2.2.4 *Algorithms for Optimal Train Control*

According to the existing literature, the energy-saving optimal train control problem was mainly solved using analytical algorithms, numerical algorithms, and evolutionary algorithms [39].

- Analytical algorithms: based on the optimal control theory such as Pontryagin's maximum principle, the necessary conditions for the optimal control can be derived. Then, the optimal sequence and the switching points of the driving regimes are obtained by solving the differential equation of the train motion under the optimal control scheme [1–3, 14]. This method can obtain the theoretically optimal solution under each specific mathematical model with different line parameters (e.g., gradients, speed limits), but it requires rigid properties of the formulated mathematical models [39].

- Numerical algorithms: involving algorithms such as dynamic programming [20], sequential quadratic programming [25], Lagrange multiplier-based algorithm [28], and nonlinear programming based on pseudospectral methods [31], have relatively less requirements for the objective function, and can make a trade-off between optimisation performance and computational time. Theoretically, this kind of solution approach can obtain the optimal solution given enough computational time [39]. Besides, a near-optimal solution can also be derived in a shorter calculation time.
- Evolutionary algorithms: by simulating some natural phenomena or processes, heuristic algorithms can repeatably search for high-quality driving strategy in the searching space. Compared with the former two kinds of methods, evolutionary algorithms, e.g., genetic algorithm [7], ant colony optimisation [18], and simulated annealing algorithm [19], have the least requirements for models in train speed profile optimisation [39]. Nevertheless, because of the random nature, most of these algorithms cannot guarantee the optimality and convergence of the solutions.

In this book, analytical algorithms for the optimal train control will be introduced in Chap. 3.

2.3 Energy-Efficient Train Timetable

In this section, the energy-efficient optimisation method of a train timetable will be briefly introduced. The energy-efficient driving strategy introduced in Sect. 2.2 is the foundation of an energy-saving train timetable to obtain the train trajectory between stops under a fixed running time. Further, the parameters of train timetable (e.g., running time between stops) can be optimised to systematically reduce the traction energy consumption in a rail network.

2.3.1 Train Timetable of Railway System

A train timetable can be visualized as a diagram of the relationship between time and space of the train operation. As shown in Fig. 2.9, it is a two-dimensional line diagram that shows the operation of each train in the segment and the dwelling of each train at the station. In the train timetable, the order of the operated trains, the arrival, departure, or passing time at each station, the inter-station running time, the dwell time at the station, and the turn around of rolling stocks are determined. Besides, the corresponding relationship among train services in the temporal dimension and the spatial dimension can be intuitively represented.

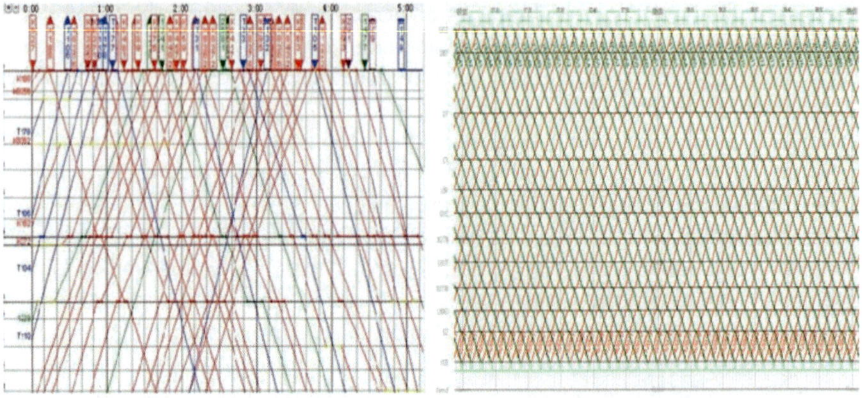

(a) Mainline railway train timetable (b) Metro train timetable

Fig. 2.9 Train timetable

In terms of representation, the abscissa in the train timetable represents the time variable and the ordinate represents the distance segmentation. The horizontal line and vertical line in the train timetable represent the centre position of each station and the time segment, respectively. The diagonal lines in the timetable represents the line of train operation, which is the profile for the planned operation of each train. Each set of diagonal lines corresponds to a specific service number for the train. The intersection points of the train profiles and the horizontal lines in the timetable represent the activity of arrival, departure, or passing.

When designing the train timetable, it is necessary to take into account the track capacity, vehicle performance, passenger demand and other factors to ensure the operation safety and efficiency. Therefore, the following main constraints need to be considered.

- The train headway should meet the minimum headway constraint determined by the line resource;
- The dwell time at each station should be longer than the minimum time for passengers to complete the necessary boarding and alighting. The dwell time is also required to be less than a maximum time to ensure the efficiency of train operation;
- The minimum inter-station running time constraint should be met according to the line resource and vehicle performance. The maximum inter-station running time is related to the passenger demand.

Besides, the train operation may be affected by disturbances, such as bad weather, equipment failure, and other factors. In these cases, train delays which affect the passenger services are likely to occur. To reduce the impact of train delays, running time supplements are usually added when scheduling the train timetable, which can be used to realize delay recovery without train timetable adjustment. After allocating the running time supplement to each segment, the actual running time between stops is composed of the minimum running time and the allocated supplement.

2.3.2 Relationship Between Train Timetable and Traction Energy

As introduced in Sect. 2.2.2, the driving strategy in each segment is directly related to the traction energy consumption in the segment. Besides, the train driving strategy is related to the parameters of train timetable. Therefore, to systematically reduce the traction energy of the whole system, the optimisation of the train timetable is also necessary. In this chapter, the relationship between the train timetable and traction energy will be introduced.

Firstly, there is a direct relationship between the inter-station running time and driving strategy. On the premise of meeting the inter-station running time constraints, there are still a variety of optional running time for trains in each segment. Normally, the traction energy consumption of train operation can be reduced by increasing the inter-station running time. The inter-station running time is composed of minimum running time and running time supplement. The minimum running time is fixed when the parameters of vehicle and route are determined. Therefore, the more the running time supplement in each segment, the better the energy-efficient effect. With a given total running time of several successive stops in a corridor, the total running time supplement for a train is fixed. As a result, how to allocating the total running time supplement to each segment to achieve the systematic energy-saving over the entire line is a key problem of train timetable optimisation [29, 30].

Another energy-saving issue will arise when considering multiple trains running in the rail network. During the train operation, train path conflicts may occur when two trains aim to occupy a shared track simultaneously. To ensure the safety of train operation, a train needs to stop running until the route ahead is free again. However, this will cause the train to restart which consume more traction energy. Therefore, it is important to avoid the route occupation conflicts when designing the energy-efficient train timetable.

Based on the above two aspects of energy-saving theory, the optimisation of energy-efficient train timetable can be divided into three levels (see Fig. 2.10). The micro level is energy-efficient train trajectories of trains, which are directly related to the traction energy consumption. The corridor timetables are located in the middle level. Each corridor timetable determines the running time between each stops and the dwell time at each stop for a train in the corridor. The macro level is network timetable, which plans the routes and times of all trains in the rail network. There are connections among the three levels when dealing with the energy-efficient train operation problem. The middle level determines the corridor constraints to be satisfied when generating the train trajectories at the micro level. Conversely, the train running trajectory determines the running time between stops. Therefore, the micro

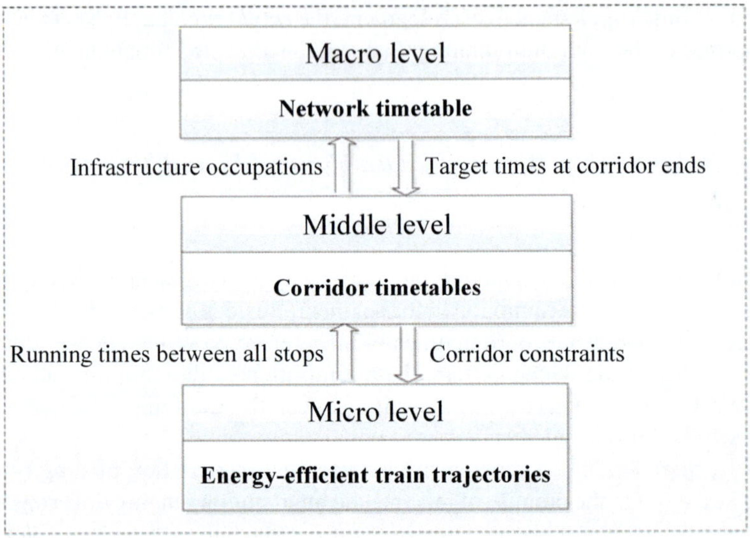

Fig. 2.10 Three levels of the energy-efficient train timetable

and middle levels need to be considered together when allocate the total running time supplement to each segment. Besides, the designing of corridor timetables should consider the total running time in a corridor determined by the network timetable. The occupation conflict in a network timetable is related to the occupation conditions in the middle level. To design a conflict-free timetable, both the middle and macro levels need to be considered.

In Chap. 4, the detailed train timetable optimisation methods will be introduced according to the relationship of levels introduced above.

2.4 Optimisation of Train Timetables for Regenerative Braking

In this section, the principle of an integrated optimisation approach will be briefly introduced to optimise the train timetable and driving strategy. The same as Sect. 2.3.2, the operation of multiple trains in multiple segments are considered in this approach. What is more, this approach also considers the reuse of feedback RBE in addition to the reduction of traction energy consumption.

2.4.1 Feedback of RBE

There are two main ways to reduce the net energy consumption by using the train-based energy-efficient strategy. Apart from decreasing the traction energy, efficient reuse of RBE is another important mean. When applying the regenerative braking mode, the motor of the train is converted into a generator during braking, which converts the kinetic energy into RBE while providing braking force. The RBE that can be recycled accounts for a large proportion of the kinetic energy (e.g., about 80% for urban rail transit). Therefore, increasing the amount of utilized RBE is crucial for the energy-saving train operation.

One of the important ways to reuse RBE is feeding back RBE to the catenary. The fed-back RBE will make the voltage of the network around the braking train(s) higher than the voltage around the traction train(s). Then, there will be current flowing from the braking train(s) to traction train(s), which means the RBE is transmitted to the traction trains and reused.

2.4.2 Relationship Between Train Operation and Feedback RBE

As introduced above, the key to make full use of the feedback RBE is to match the time when adjacent trains take traction/braking regime. Therefore, it is necessary to collaborate the operation of multiple trains. During the operation, the driving regime of trains is influenced by both the driving strategy and the train timetable. Specifically, the driving strategy determines the driving regimes applied in each moment during the inter-station operation. The train timetable determines when the train departs from the station and starts an inter-station running. Therefore, the integrated optimisation approach plays an important role to efficiently reused the RBE. In this approach, the driving strategy and the train timetable can be simultaneously optimised, making trains cooperate to apply traction/braking regimes. Figure 2.11 shows the relationship between the spatio-temporal position of the train and the energy consumption under two conditions. The departure time of train i' is different in these two conditions. When the RBE is generated by the braking train, if there is no train nearby to implement the acceleration regime, the generated energy will increase the voltage of catenary. When the catenary voltage rises to a threshold, the over-voltage protection will be taken and the generated RBE will be consumed by the heating resistance. When the braking regime and acceleration regime are matched successfully and the two trains are close to each other, the RBE can be reused by the traction train.

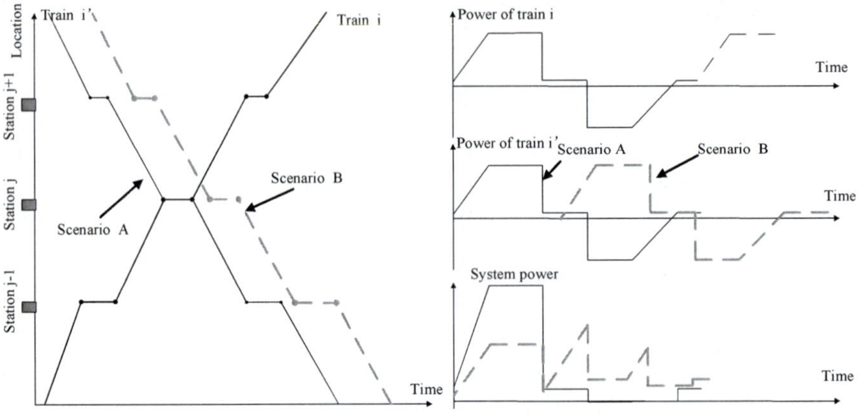

Fig. 2.11 Increasing the reused of RBE by matching the regimes [1]

2.4.3 Categories of Integrated Optimisation Methods

There are two main integrated optimisation methods to minimise the net energy consumption.

The first method is a two-level method, in which the driving strategy is optimised in the lower level and the train timetable is optimised in the higher level [26, 32, 33, 35, 38]. In this method, the running time between stations is the bridge connecting two levels. In the macro level, the arrival time and departure time are adjusted to reduce the net energy consumption and determine the running time between stations. In the micro level, the driving strategy between stations with the minimum traction energy consumption is calculated according to the running time determined in the macro level. As a result, the driving strategy between stations is unique when running time between stations is determined. However, the driving strategy is optimised without considering the reuse of RBE in the two-level method, while the amount of reused RBE is related to the synchronization of the braking/traction regimes. As a result, the impact of both the train timetable and the driving strategy on the reuse of RBE need to be considered to increase the amount of reused RBE.

The second integrated optimisation method is to synchronously optimise the driving strategy and the train timetable. In this method, In this method, the efficient utilization of RBE can also be considered when optimising the driving strategy between stations, so as to achieve better systematic energy saving. For example, the switching points of driving regimes and the train timetable are jointly optimised in a synchronous method [21, 22]. In this method, the time to apply braking and traction regimes in the segment can be matched as well as synchronizing the arrival and departure time.

In Chap. 5, both the two-stage method and the synchronous method will be detailed.

2.5 Energy-Efficient Driving Considering ESSs

In this section, another approach to increase the reused RBE will be introduced with the consideration of ESSs. In this approach, the driving strategy of a single train on the whole line can be optimised according to the power characteristics of ESSs.

2.5.1 ESS in Railway System

Apart from feeding back RBE, another way to reused RBE is to use ESSs. Regarding the technologies available for ESSs, Electrochemical Double Layer Capacitors (EDLC), batteries and flywheels are the most options so far.

An ESS can be installed either on the vehicles (on-board ESSs) or at specific points along the track (track-side ESSs). By using on-board ESSs, the produced RBE can be temporarily stored on the train. The stored RBE can be used by the on-board ancillary devices such as lighting, or the next acceleration regime of train operation. In general, an on-board ESS has a higher efficiency owing to the absence of line losses. However, large space on the vehicles and a considerable increase of weight is required by on-board ESSs, which increase the cost and may hinder their installation in existing rolling stock.

As for the track-side ESSs, it can collect the RBE from braking trains and release the stored energy if other train uses the acceleration regime. Compared with an on-board ESS, the track-side ESSs present fewer weight and space restrictions. Besides, the installation and maintenance of track-side ESSs do not influence other services. However, track-side ESSs are generally less efficient due to transmission losses taking place in the network.

Both on-board and track-side ESSs can lead to considerable energy savings in railway systems (typically between 15 and 30% in urban rail transit [12]). moreover, ESSs may contribute to stabilising the network voltage and to shaving the power consumption peaks.

2.5.2 Control of ESSs During Train Operation

In the control of on-board ESSs, the parameters including the vehicle speed, State of Charge (SoC), requested traction power and network voltage should be considered. On one hand, control systems should ensure that ESSs are charged enough to power the vehicle during accelerations. On the other hand, ESSs should remain completely discharged when the train speed is high to accept the highest amount of energy when breaking or dwelling [9, 16].

The track-side ESSs can absorb the RBE that cannot be used directly in the system. The ESSs deliver the stored energy when it is required for the traction of any train in its electric section. The charge and discharge processes require an electronic controller that generally operates as a function of the voltage on the line: when an overvoltage takes place as a result of any braking process, ESSs operate in "charging" mode to absorb the excess RBE on the line; in turn, when a voltage drop is detected, ESSs will deliver the stored energy in order to keep the threshold value on the network [17, 23].

Chapter 6 will introduce the energy-efficient driving methods considering both the on-board ESSs and the track-side ESSs.

2.6 Substation-Based Energy-Efficient Strategy for Train Operation

The train-based energy-efficient strategy is introduced in Sect. 2.2–2.4. In this section, the substation-based energy-efficient strategy will be expounded with the combination of the TPSS characteristics, to minimise the energy consumption offer the by the substations.

2.6.1 Advantage of Substation-Based Energy-Efficient Strategy

The train-based energy-efficient strategy introduced in the above section aims to reduce the total net energy consumption of all the operated trains. In the train-based method, the traction energy of all trains is summed up to get the total traction energy consumption. Besides, a simplified formula is applied to calculate the reused RBE by considering the factors including the distance between trains, train regimes, etc. However, there is still a gap between the net energy consumption calculated by the train-based method and the actual energy consumption of the Traction Power Supply System (TPSS). Firstly, the transmission process of the RBE is not only related to the state of operated trains, but also affected by the real-time state of the TPSS, e.g., voltage and current. Therefore, the amount of RBE cannot be precisely calculated if only the train operation is considered. Besides, when the electrical energy provided by the substation is transmitted to the train through the traction network, the transmission loss will be generated. The transmission loss is not only related to the position and the power of trains, but also influenced by the parameters of the traction network. Therefore, the train-based method cannot accurately calculate the systematic energy consumed by train operation. The substation-based energy-efficient strategy is necessary, in which both the state of train operation and the state of TPSS are considered to minimise the total energy consumption offered by the TPSS.

2.6.2 *Principle of Substation-Based Energy-Efficient Strategy*

The optimisation of substation-based energy-efficient strategy should be integrated considering the train state and TPSS state. On one hand, the train state at each instant is determined by both the train timetable and the driving strategy. On the other hand, the train state is also the data basis to calculate the energy consumption of the TPSS by considering the dynamic TPSS state. Therefore, the instantaneous state of a train is the link between train timetable/driving strategy and the TPSS energy consumption.

In the energy-saving optimisation, the TPSS energy consumption calculation model should be established first, to calculate the energy consumption according to the real-time train state and TPSS parameters. Then, the train operation model should be developed to get the real-time train state based on the train timetable and driving strategy. Finally, the above two models are combined to solve the substation-based energy-efficient strategy [8, 27, 40].

2.7 Conclusion

In this chapter, the train-based and the substation-based energy-efficient strategies were introduced separately. The train-based energy-efficient strategy was divided into four methods according to the objective function and decision variables. In each method, the relationship between the train operation and the amount of energy consumption was introduced. These four methods will be expounded in Chaps. 3–6. Then, the advantage and the principle of substation-based energy-efficient strategy which will be explained in Chap. 7.

References

1. Albrecht T (2014) Energy-efficient railway operation. In: Hansen IA, Pachl J (eds) Railway timetabling and operations. Eurailpress, pp 91–116
2. Albrecht A, Howlett P, Pudney P, Vu X, Zhou P (2016) The key principles of optimal train control-part 1: formulation of the model, strategies of optimal type, evolutionary lines, location of optimal switching points. Transp Res Part B Methodol 94:482–508
3. Albrecht A, Howlett P, Pudney P, Vu X, Zhou P (2016) The key principles of optimal train control-part 2: existence of an optimal strategy, the local energy minimization principle, uniqueness, computational techniques. Transp Res Part B Methodol 94:509–538
4. BSI (2009) BS EN 12299:2009: railway applications—ride comfort for passengers-measurement and evaluation
5. Cacchiani V, Caprara A, Galli L, Kroon L, Maróti G, Toth P (2012) Railway rolling stock planning: robustness against large disruptions. Transp Sci 46(2):217–232
6. Cacchiani V, Huisman D, Kidd M, Kroon L, Toth P, Veelenturf L, Wagenaar J (2014) An overview of recovery models and algorithms for real-time railway rescheduling. Transp Res Part B Methodol 63:15–37

7. Chang CS, Sim SS (1997) Optimising train movements through coast control using genetic algorithms. IEE Proc Electr Power Appl 144(1):65–73
8. Chen M, Feng X, Wang Q, Sun P (2021) Cooperative eco-driving of multi-train under dc traction network. IEEE Trans Transp Electrif 7(3):1805–1821
9. Domínguez M, Fernández-Cardador A, Cucala AP, Pecharromán R (2012) Energy savings in metropolitan railway substations through regenerative energy recovery and optimal design of ATO speed profiles. IEEE Trans Autom Sci Eng 9(3):496–504
10. Dong H, Ning B, Cai B, Hou Z (2010) Automatic train control system development and simulation for high-speed railways. IEEE Circ Syst Mag 10(2):6–18
11. ERA UNISIG (2018) ATO over ETCS system requirements specification. REF: SUBSET 125, ISSUE: 0.1.0
12. González-Gil A, Palacin R, Batty P (2013) Sustainable urban rail systems: strategies and technologies for optimal management of regenerative braking energy. Energy Convers Manage 75:374–388
13. González-Gil A, Palacin R, Batty P, Powell JP (2014) A systems approach to reduce urban rail energy consumption. Energy Convers Manage 80:509–524
14. Howlett PG, Milroy IP, Pudney PJ (1994) Energy-efficient train control. Control Eng Practice 2(2):193–200
15. Huang Y, Yang L, Tang T, Cao F, Gao Z (2016) Saving energy and improving service quality: bicriteria train scheduling in urban rail transit system. IEEE Trans Intell Transp Syst 17(12):3364–3379
16. Huang Y, Yang L, Tang T, Gao Z, Cao F, Li K (2018) Train speed profile optimization with on-board energy storage devices: a dynamic programming based approach. Comput Industr Eng 126:149–164
17. Kampeerawat W, Koseki T (2017) A strategy for utilization of regenerative energy in urban railway system by application of smart train scheduling and wayside energy storage system. Energy Procedia 138:795–800
18. Ke BR, Lin CL, Yang CC (2012) Optimisation of train energy-efficient operation for mass rapid transit systems. IET Intell Transp Syst 6(1):58–66
19. Kim K, Chien SIJ (2011) Optimal train operation for minimum energy consumption considering track alignment, speed limit, and schedule adherence. J Transp Eng 137(9):665–674
20. Ko H, Koseki T, Miyatake M (2004) Application of dynamic programming to the optimization of the running profile of a train. WIT Trans Built Environ 74
21. Li X, Lo HK (2014) An energy-efficient scheduling and speed control approach for metro rail operations. Transp Res Part B Methodol 64:73–89
22. Li X, Lo HK (2014) Energy minimization in dynamic train scheduling and control for metro rail operations. Transp Res Part B Methodol 70:269–284
23. Luan X, Wang Y, De Schutter B, Meng L, Lodewijks G (2018) Integration of real-time traffic management and train control for rail networks-part 2: extensions towards energy-efficient train operations. Transp Res Part B Methodol 115:72–94
24. Matsumo M (2005) ISADS 2005. IEEE 2005:599–606
25. Miyatake M, Ko H (2010) Optimization of train speed profile for minimum energy consumption. IEEJ Trans Electr Electron Eng 5(3):263–269
26. Ning J, Zhou Y, Long F, Tao X (2018) A synergistic energy-efficient planning approach for urban rail transit operations. Energy 151:854–863
27. Pan Z, Chen M, Lu S, Tian Z, Liu Y (2020) Integrated timetable optimization for minimum total energy consumption of an AC railway system. IEEE Trans Vehic Technol 69(4):3641–3653
28. Rodrigo E, Tapia S, Mera JM, Soler M (2013) Optimizing electric rail energy consumption using the Lagrange multiplier technique. J Transp Eng 139(3):321–329
29. Scheepmaker GM, Goverde RMP (2015) The interplay between energy-efficient train control and scheduled running time supplements. J Rail Transp Plann Manage 5(4):225–239
30. Scheepmaker GM, Pudney PJ, Albrecht AR, Goverde RMP, Howlett PG (2020) Optimal running time supplement distribution in train schedules for energy-efficient train control. J Rail Transp Plann Manage 14:100180

31. Goverde RMP, Scheepmaker GM, Wang P (2021) Pseudospectral optimal train control. Euro J Oper Res 292(1):353–375
32. Su S, Li X, Tang T, Gao Z (2013) A subway train timetable optimization approach based on energy-efficient operation strategy. IEEE Trans Intell Transp Syst 14(2):883–893
33. Su S, Wang X, Cao Y, Yin J (2019) An energy-efficient train operation approach by integrating the metro timetabling and eco-driving. IEEE Trans Intell Transp Syst 21(10):4252–4268
34. Tang T, Huang LJ (2003) A survey of control algorithm for automatic train operation. J China Railway Soc 25(2):98–102
35. Tian Z, Weston P, Zhao N, Hillmansen S, Roberts C (2017) System energy optimisation strategies for metros with regeneration. Transp Res Part C Emerg Technol 75:120–135
36. Wang Z, Quaglietta E, Bartholomeus MGP, Goverde RMP (2022) Assessment of architectures for automatic train operation driving functions. J Rail Transp Plann Manage 24:100352
37. Yasunobu S (1985) Automatic train operation by predictive fuzzy control. Industr Appl Fuzzy Control 1985:1–18
38. Yin J, Yang L, Tang T, Gao Z, Ran B (2017) Dynamic passenger demand oriented metro train scheduling with energy-efficiency and waiting time minimization: mixed-integer linear programming approaches. Transp Res Part B Methodol 97:182–213
39. Yin J, Tang T, Yang L, Xun J, Huang Y, Gao Z (2017) Research and development of automatic train operation for railway transportation systems: a survey. Transp Res Part C Emerg Technol 85:548–572
40. Zhao N, Roberts C, Hillmansen S, Tian Z, Weston P, Chen L (2017) An integrated metro operation optimization to minimize energy consumption. Transp Res Part C Emerg Technol 75:168–182
41. Zhou X, Zhong M (2007) Single-track train timetabling with guaranteed optimality: branch-and-bound algorithms with enhanced lower bounds. Transp Res Part B Methodol 41(3):320–341
42. Zimmermann A, Hommel G (2005) Towards modeling and evaluation of ETCS real-time communication and operation. J Syst Softw 77(1):47–54

Chapter 3
Energy-Efficient Driving for a Single Train

Peter Pudney

3.1 Introduction

This chapter describes the mathematical modelling and analysis required to determine the most energy-efficient way to drive a single train so that it arrives on time at every timing point along its journey. It also describes how the theory of optimal train control has been used to develop practical driving advice systems.

Work on energy-efficient train control began in the 1970s and 1980s, often motivated by the high cost of diesel fuel. Today, reducing greenhouse gas emissions and reliance on fossil fuels is crucial. Clean energy is not necessarily cheap energy, so reducing energy use is still important.

Another useful outcome of driving advice systems is precise adherence to schedules. For many railways, this is more important than energy use. With a well-designed schedule, precise timekeeping will reduce train delays due to encounters with restrictive signals.

The chapter is based on the work of the Scheduling and Control Group at the University of South Australia. The predecessor to this group was formed at the South Australian Institute of Technology in the early 1980s to develop a driving advice system for suburban trains. This system, called Metromiser, was deployed on the suburban rail network in Adelaide, South Australia, in the mid 1980s. Starting in the late 1980s, the group's focus moved to developing the theory of optimal train control, and developing practical driving advice systems that could be used for long-haul journeys. Commercial products based on this work were developed by TTG Transportation Technology (now part of Trapeze Rail), and driving advice systems have been deployed on freight, passenger and high-speed railways around the world.

The example driving profiles in this chapter were calculated using the same software that is used on trains to calculate driving advice.

P. Pudney (✉)
University of South Australia, Adelaide, Australia
e-mail: Peter.Pudney@unisa.edu.au

S. Su et al., *Energy-Efficient Train Operation*, Lecture Notes in Mobility,
https://doi.org/10.1007/978-3-031-34656-9_3

Other significant work on optimal train control is described in the review article by Scheepmaker et al. [13].

Industry uses the terms 'driver advice system' or 'driver advisory system' (DAS) to refer to systems that provide a driver with real-time driving advice to keep a train on time and minimise fuel or energy use. However, it is important to recognise that driving advice can also be used without driver intervention, as part of a cruise control or automatic train operation (ATO) system.

This chapter discusses the mathematical models used to predict the motion of a train and its energy use (Sect. 3.2), and then describes the formulation and solution of an optimal control problem that minimises the energy required to complete a journey on time, starting with a simple journey on level track with no speed limits, then progressing to journeys with gradients and speed limits (Sect. 3.3). The next sections discuss the relationship between journey duration and energy (Sect. 3.4), the impact of regenerative braking on optimal control strategies (Sect. 3.5), how having more power available enables more efficient driving (Sect. 3.6), and how to handle intermediate time constraints and timing windows (Sect. 3.7). A final section describes how the optimisation methods can be used to provide real-time driving advice on a train (Sect. 3.8).

3.2 Modelling the Motion of a Train

The motion of a train along a track is controlled by varying control inputs to the traction and braking systems. This in turn determines the acceleration of the train, the speed of the train, the time that the train reaches specified locations, and the fuel or energy consumed by the train. Our aim is to determine a control strategy that will arrive at timing locations on time and minimise the fuel or energy used for the journey.

The main forces acting on a train are:

- the force of the wheels against the track, produced by the traction and braking systems
- forces due to gravity and the gradient of the track
- rolling resistance and aerodynamic drag forces that oppose the motion of the train
- forces due to the curvature of the track.

There are some significant forces that are impractical to model. For example, head winds and side winds can have a significant impact on a long train, but it is not possible to predict these forces for the entirety of a long journey. For a practical driving advice system, unknown or changing future conditions can be handled using an adaptive, 'model predictive control' approach. At any instant the optimal control is calculated using the best available knowledge. If the train strays from the predicted path then a new optimal path is created starting from the actual state of the train. For this to work in real time, calculation of an optimal control strategy for the remainder of the journey must be fast.

There are other factors that are usually ignored when calculating long-term energy-efficient driving strategies:

- In-train forces are important on long, heavy trains, but do not have a large impact on the long-term driving strategy or on energy consumption. Drivers wishing to keep a long train stretched or compressed can, for example, use low traction or braking settings instead of coasting, with little impact on the overall strategy or energy use.
- Traction system constraints can limit how quickly the traction system can change its output. The details differ across different traction systems. The impact on the overall driving strategy and energy consumption is small.
- Air braking systems for long trains vary, and can be difficult to model. Instead, we assume simple but conservative braking performance curves that we know the train can follow.

3.2.1 Tractive Effort

Most trains have electric traction motors powered by either an external electrical supply or by an on-board generator. In any case, the maximum tractive effort that can be supplied at the wheels varies with the speed of the train, and generally decreases as speed increases. Figure 3.1 shows tractive effort and braking effort curves for an example 2200 kW diesel-electric locomotive. Figure 3.2 shows tractive effort and braking effort curves for an example 7000 kW high-speed electric passenger train. The braking curves are explained in the next section.

Both tractive effort curves have two distinct regions, typical of tractive effort curves. At lower speeds tractive effort is limited by adhesion or by traction motor current; at higher speeds the tractive effort is constrained by power limits. Traction control systems will also influence the shape of the tractive effort curve. For both these examples the tractive curve is defined by tabulated data from the train manufacturer.

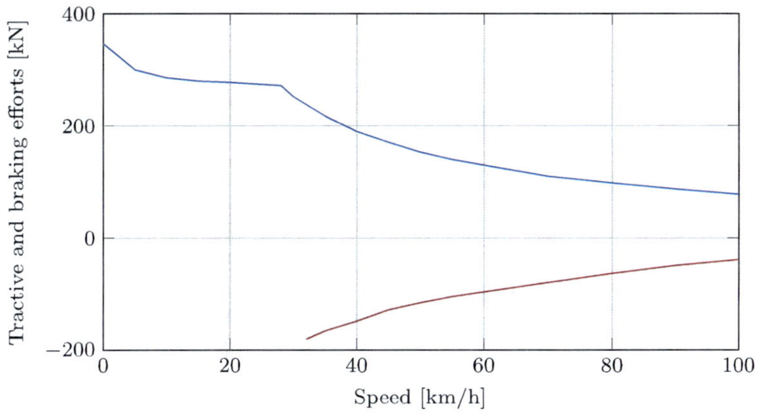

Fig. 3.1 Tractive effort (top) and braking effort (bottom) for a 2200 kW diesel-electric locomotive

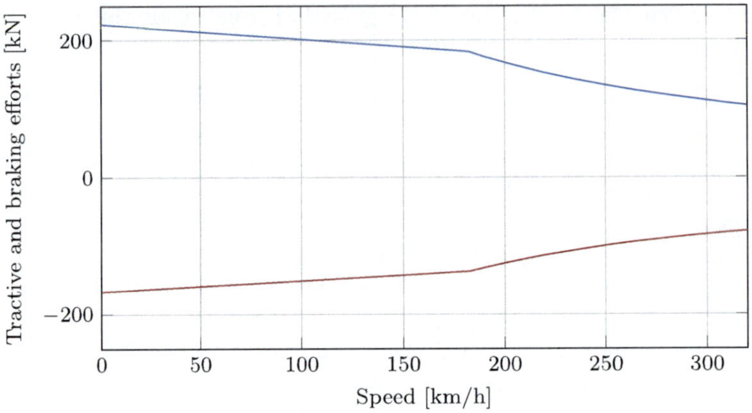

Fig. 3.2 Tractive effort (top) and braking effort (bottom) for a high-speed electric passenger train

If detailed tractive effort data is not available, tractive effort can be approximated by the formula

$$T(v) = \min\{T_{\max}, P/v\}$$

where $T_{\max}$ is the maximum tractive effort and P is the maximum tractive power. Tractive effort is often limited by wheel adhesion, so $T_{\max}$ can be approximated by

$$T_{\max} = kmg$$

where g is acceleration due to gravity, mg is the downward force on the driven wheels and k is a coefficient of friction, typically around 0.35 for heavy haul trains.

3.2.2 *Braking Effort*

Trains usually have two braking systems:

- air brakes, which use compressed air to control the pressure of brake pads onto brake discs attached to the wheels
- an electric braking system, which uses the electric traction motors to generate electrical energy that is either dissipated as heat using resistors on the train (dynamic braking) or else fed back to the external electrical supply system or stored in an on-board energy storage system (regenerative braking).

Electric braking cannot provide braking effort at very low speeds, so a blend of air and electric braking is used at low speeds. The braking effort curve shown in Fig. 3.1 is the electric braking effort provided by the locomotive for speeds greater than 30 km/h. At lower speeds the electric braking effort reduces, so air braking must be used. The braking effort curve in Fig. 3.2 is controlled at speeds below 180 km/h to control the deceleration of the train.

Electric braking is often easier to control than air brakes. For many trains, preferred braking practice is to use electric braking as much as possible, even though air brakes can provide greater braking effort. Some passenger trains have braking control systems that give constant braking effort over a wide range of speeds.

For modelling energy-efficient driving, it is usually assumed that the train has a single maximum braking effort curve that represents good driving practice, and is implemented as electric braking at high speed blended with air braking at low speed.

3.2.3 Resistance Forces

Resistance forces are forces that arise because of the motion of the train, and oppose the motion of the train. They include:

- rolling resistance, due primarily to the deformation of the wheel and rail as the wheel rolls along the track
- bearing resistance
- aerodynamic drag.

The key characteristic of rolling resistance is that resistance R is an increasing function of speed v, with increasing derivative R'.

Early work modelled resistance force as a quadratic in the speed of the train,

$$R(v) = r_0 + r_1 v + r_2 v^2$$

where the coefficients r_0, r_1 and r_2 are all non-negative and depend on factors including the train mass, number of axles, train length and the aerodynamic efficiency of the train. Davis [6] developed formulas for estimating these coefficients from these more basic characteristics of the train. More recent work by Lukaszewicz [9] gives formulas for more modern European rolling stock.

It is important to recognise that R is not directly proportional to the mass of the train. While the constant and linear term coefficients r_0 and r_1 are generally proportional to mass, the quadratic term coefficient r_2 is mostly dependent on aerodynamic drag, which is independent of mass.

3.2.4 Gradient Forces

The gradient of a track is

$$\gamma = \frac{\Delta y}{\Delta x}$$

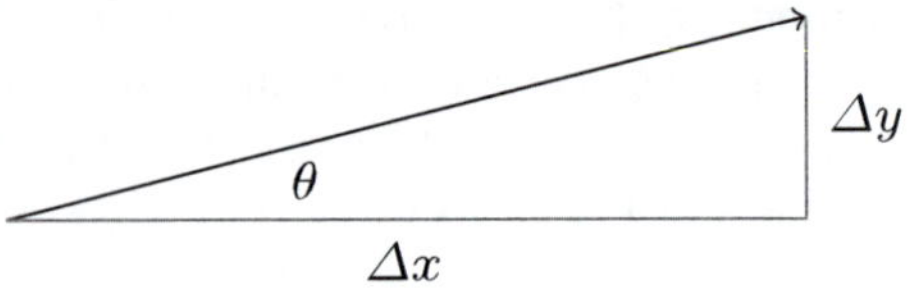

Fig. 3.3 Gradient of a track

where the 'run' Δx is a horizontal distance in the direction of travel and the 'rise' Δy is the corresponding increase in elevation. The angle of slope is given by

$$\theta = \arctan \gamma$$

(see Fig. 3.3).

The force acting on a train on a track with constant gradient γ is

$$G = -mg \sin \theta$$

where m is the mass of the train and $g \approx 9.8\ \mathrm{ms}^{-2}$ is the acceleration due to gravity. Gradient force G is negative for inclines and positive for declines.

A gradient such as $\gamma = 0.02$ can also be expressed as 2 or 20% or '1 in 50'.

For the small gradients typical of railways, $\sin \theta \approx \gamma$ and so $G \approx mg\gamma$.

Long trains may extend over a changing gradient. How this is handled is described in Sect. 3.2.6.

3.2.5 *Track Curvature Forces*

The force on a train due to track curvature is usually small compared to traction, braking and gradient forces. It is usually modelled as depending on the curvature of the track and not on the speed of the train.

The curvature of a track at a given point on the track is defined as

$$\kappa = 1/r$$

where r is the radius of the circular arc that best approximates the track at that point. Curvature can be signed, to indicate the direction of curvature.

Track curvature is often specified as a sequence of straight (tangent) track segments and curved segments with given radii.

Lukaszewicz [9] gives the following empirical formula for the force due to curvature at distance x along a track:

$$C(x) = 0.455\, mg \frac{|\kappa(x)|}{1 - 55\, |\kappa(x)|}$$

where m is the mass of the train, $g \approx 9.8 \text{ ms}^{-2}$ is the acceleration due to gravity, and $\kappa(x)$ is the curvature of the track at distance x.

3.2.6 *Transformed Track Forces for Long Trains*

If a train is long enough that the gradient or curvature of the track varies beneath the train then we need to take into account the fact that the gradient and curvature forces will vary along the length of the train.

If the front of the train is at location x and the train has length l then the total gradient force on the train is

$$\hat{G}(x) = -\int_0^l \rho(s) g \sin\theta(x - s)\, ds$$

where $\rho(s)$ is the mass per unit length of the train at distance s from the front of the train, and $\theta(x - s)$ is the angle of slope of the track at location $x - s$. The total gradient force will be different to the gradient force at the front of the train if the gradient is non-constant, even if the mass is distributed uniformly along the train—the integral over the length of the train essentially smooths and shifts the gradient force.

The motion of a point mass train with the transformed gradient force $\hat{G}$ is equivalent to the motion of the original long train with the original gradient force G. This allows us to treat all trains as point masses, by first transforming the gradient force.

A similar transformation can be done with curvature force.

In practice the precise mass distribution of a train may not be known, particularly for freight trains. It is usually accurate enough to assume that the mass of the locomotives is distributed uniformly along their length and the trailing mass is spread uniformly along the remainder of the train.

The remainder of this chapter assumes a point mass train, and uses the simpler notations $G(x)$ and $C(x)$ to denote transformed gradient force and transformed curvature force at distance x along the route.

3.2.7 *Equations of Motion*

A train can be treated as a point mass if the track forces are first transformed to take into account the length and mass distribution of the train (Sect. 3.2.6). The driver controls the traction and braking effort u, with $u \in [0, T(v)]$ for traction and $u \in [B(v), 0]$ for braking. The equation of motion of the train is then given by

$$m\frac{dv}{dt} + I\frac{d\omega}{dt} = u - R(v) + G(x) - C(x)$$

where

- m is the mass of the train
- v is speed of the train along the route
- t is time
- I is the moment of inertia of rotating parts of the train such as the wheels and traction motor rotors
- ω is the angular velocity of rotating parts
- u is the controlled tractive force ($u \geq 0$) or braking force ($u \leq 0$)
- $R(v) \geq 0$ is the total resistance force acting against the motion of the train at speed v
- $G(x)$ is the gradient force acting on the train at distance x along the route
- $C(x) \geq 0$ is the track curvature force acting against the motion of the train at distance x along the route.

This equation of motion can be simplified by noting that the angular velocity of rotating parts is typically proportional to the speed of the train, so we can write

$$\hat{m}\frac{dv}{dt} = u - R(v) + G(x) - C(x)$$

where $\hat{m}$ is the *effective mass* of the train. It is usually assumed that $\hat{m} = \rho m$ where ρ is a *rotating mass factor*, typically in the range [1.0, 1.1].

Dividing both sides by the effective mass gives an equation for the acceleration of the train:

$$\frac{dv}{dt} = \frac{1}{\hat{m}}\left[u - R(v) + G(x) - C(x)\right]. \tag{3.1}$$

This version of the equation of motion has time as the independent variable. But gradient and curve forces depend on distance, not time, so it is convenient to use distance as the independent variable and write

$$\frac{dv}{dx} = \frac{1}{v}\frac{dv}{dt} = \frac{1}{\hat{m}v}\left[u - R(v) + G(x) - C(x)\right]. \tag{3.2}$$

This formulation introduces the problem that dv/dx is undefined when $v = 0$. In practice, we can work around this problem by starting calculations at some small speed $v_0 = \varepsilon$; the example journeys in this chapter start at $v_0 = 1\ \text{ms}^{-1}$.

3.2.8 Energy Use

The electrical energy consumption rate of an electric train or the fuel consumption rate of a diesel-electric train depends on both the tractive power being generated and on the efficiency of the traction system. In general, energy consumption rate and fuel consumption rate both increase with tractive power. This means that energy or fuel use can be minimised by minimising the mechanical work done by the traction system, without modelling the precise relationship between electrical energy consumption or fuel consumption and mechanical power.

The mechanical work done by the traction system in moving the train from $x = A$ to $x = B$ is traction force times over distance, and so

$$W_{\mathrm{t}} = \int_A^B u_{\mathrm{t}}\,dx$$

where

$$u_{\mathrm{t}} = \frac{u + |u|}{2} = \begin{cases} u & u \geq 0 \\ 0 & u \leq 0. \end{cases}$$

On electric trains, or on hybrid trains with on-board energy storage, it is possible to recover some of the electrical braking energy of the train. The mechanical work done by the braking system during a journey is

$$W_{\mathrm{b}} = \int_A^B u_{\mathrm{b}}\,dx$$

where

$$u_{\mathrm{b}} = \frac{u - |u|}{2} = \begin{cases} 0 & u \geq 0 \\ u & u \leq 0. \end{cases}$$

The amount of energy that can be recovered from electric braking depends on several factors. First, energy recovery depends on the efficiency of the conversion from mechanical braking power to electrical power by the traction system. Then, for electric trains with an external supply, energy recovery depends on the receptivity of the external system. For some electrical systems, power can be transferred via the local substation back to the wider electricity grid. For other systems, power can be transferred from a braking train only if there is another train using power on the same electrical section. For hybrid trains, braking energy can be stored only if the on-board energy store is not full.

For simplicity, we will assume that the energy recovered is a fixed proportion $\rho \in [0, 1)$ of the mechanical work done by the electrical braking system. The energy for the journey is then modelled as

$$E = \int_A^B u_{\mathrm{t}} + \rho u_{\mathrm{b}}\,dx. \tag{3.3}$$

3.3 Minimising Energy with On-Time Arrival

This section gives an overview of the theory used to derive the optimal control strategy for a journey. Full details and further references are available in the articles by Albrecht et al. [4, 5], and in the articles referenced by Scheepmaker et al. [13].

3.3.1 *Formulating an Optimal Control Problem*

The simplest type of train journey starts at distance x_0 along a route, then has stops at distances $x_0 < x_1 < x_2 < \cdots < x_n$. If each stop i has a desired arrival time t_i then each journey segment between a pair of adjacent stops can be treated independently (Sect. 3.7 discusses the more general case where arrival times are not fixed).

The key problem is to determine the control function u that will drive a train starting from distance $x = x_0$ at time $t = t_0$ and speed $v_0 = 0$ to arrive at distance $x = x_1$ at time $t = t_1$ and speed $v_1 = 0$ with minimum energy consumption

$$E = \int_{x_0}^{x_1} u_\mathrm{t} + \rho u_\mathrm{b} \, dx .$$

The train system can be viewed as a dynamical system with independent variable x and control u. The state of the system is (t, v), with state equations

$$\begin{aligned} \frac{dt}{dx} &= \frac{1}{v} \\ \frac{dv}{dx} &= \frac{1}{\hat{m} v} \left[u - R(v) + G(x) - C(x) \right] . \end{aligned}$$

The initial state is (t_0, v_0) and the final state is (t_1, v_1).

Initially, we will consider journeys without train or track speed limits.

3.3.2 *Pontryagin's Principle*

Pontryagin's principle says that necessary conditions for an optimal control u^* can be found by forming a control Hamiltonian H and then selecting u^* to maximise H. The control Hamiltonian for the train control problem is

$$H = -[u_\mathrm{t} + \rho u_\mathrm{b}] + \frac{\mu_1}{v} + \frac{\mu_2}{\hat{m} v} \left[u - R(v) + G(x) - C(x) \right] \tag{3.4}$$

where μ_1 and μ_2 are continuous adjoint variables with

$$\frac{d\mu_1}{dx} = -\frac{\partial H}{\partial t} = 0 \tag{3.5}$$

$$\frac{d\mu_2}{dx} = -\frac{\partial H}{\partial v} = \frac{\mu_1}{v^2} + \frac{\mu_2}{\hat{m}v^2}[u - R(v) + G(x) - C(x)] + \frac{\mu_2}{\hat{m}v}R'(v). \tag{3.6}$$

Defining a new adjoint variable[1]

$$\lambda = \frac{\mu_2}{\hat{m}v} \tag{3.7}$$

and grouping the Hamiltonian terms that depend on the control u gives a simplified Hamiltonian

$$H = \begin{cases} [\lambda - 1]u + \cdots & u \geq 0 \\ [\lambda - \rho]u + \cdots & u \leq 0 \end{cases}$$

where the ignored terms do not depend on u. The adjoint variable λ is not defined when $v = 0$, but this is not a problem since we can avoid this situation by starting journey calculations at $v_0 = \varepsilon$.

3.3.3 *Optimal Control Modes*

The optimal control u^* maximises the Hamiltonian. There are five cases to consider, depending on the value of λ:

1. When $\lambda > 1$ the Hamiltonian is maximised by making u as large as possible, which means setting $u^* = T(v)$. This gives a driving mode corresponding to maximum tractive effort.
2. When $\lambda = 1$ the Hamiltonian is maximised by setting u to any non-negative value, which means setting $u^* \in [0, T(v)]$. More analysis is required to determine how the train should be driven when $\lambda = 1$.
3. When $\rho < \lambda < 1$ the Hamiltonian is maximised by setting $u^* = 0$. This gives a driving mode corresponding to coasting, with no traction and no braking.
4. When $\lambda = \rho$ the Hamiltonian is maximised by setting u to any non-positive value, which means setting $u^* \in [B(v), 0]$. More analysis is required to determine how the train should be driven when $\lambda = \rho$.
5. When $\lambda < \rho$ the Hamiltonian is maximised by making u as negative as possible, which means setting $u^* = B(v)$. This gives a driving mode corresponding to maximum braking.

[1] Albrecht et al. [4] also define $\eta = \lambda - 1$ and $\zeta = \lambda - \rho$.

From Eq. (3.7), the singular control mode $\lambda = 1$ corresponds to $\mu_2 = \hat{m}v$. This mode can can be maintained on a non-trivial interval only if $\mu_2' = \hat{m}v'$, where μ_2' and v' are derivatives with respect to distance. Substituting Eqs. (3.6) and (3.2) and $\mu_2 = \hat{m}v$ gives the condition

$$\mu_1 + v^2 R'(v) = 0$$

which has only one solution $v = V > 0$ with $\mu_1 < 0$ since the function

$$\psi(v) = v^2 R'(v)$$

is positive and strictly increasing. Since μ_1 is constant throughout the journey, the driving mode when $\lambda = 1$ for a non-trivial interval is always holding the constant speed V using tractive effort $u^* \in [0, T(v)]$. The value of μ_1 for the entire journey is

$$\mu_1 = -\psi(V). \tag{3.8}$$

The driving mode with $\lambda = \rho$ on a non-trivial interval requires $\mu_2' = \rho\hat{m}v'$, which gives the condition

$$\mu_1 + \rho\psi(v) = 0.$$

This also has only one solution, constant speed $W \geq V$, with partial regenerative braking $u^* \in [B(v), 0]$. This mode can be used only on declines steep enough that partial regenerative braking at speed W will not slow the train.

So an optimal journey has only five driving modes:

1. *power*, using maximum tractive effort
2. *hold*, using partial tractive effort to maintain a constant speed V
3. *coast*, with no tractive effort and no braking effort
4. *regen* (if $\rho > 0$), using partial regenerative braking effort to maintain a constant speed $W \geq V$
5. *brake*, using maximum braking effort.

3.3.4 Transitions Between Modes

Further analysis is required to determine how these modes can be pieced together to form an optimal journey. Continuity requirements for both speed v and the adjoint variable μ_2 (and hence λ) give conditions for transitions between modes, as shown in Table 3.1.

The behaviour of λ within the power, coast and brake modes depends on the track gradient and curvature, and also determines what sequences of optimal control modes are possible.

Table 3.1 Conditions for transitions between driving modes

Transition	Conditions	
Power → hold	$v = V$	$\lambda = 1$
Power → coast		$\lambda = 1$
Hold → power	$v = V$	$\lambda = 1$
Hold → coast	$v = V$	$\lambda = 1$
Coast → power		$\lambda = 1$
Coast → hold	$v = V$	$\lambda = 1$
Coast → regen	$v = W$	$\lambda = \rho$
Coast → brake		$\lambda = \rho$
Regen → coast	$v = W$	$\lambda = \rho$
Regen → brake	$v = W$	$\lambda = \rho$
Brake → coast		$\lambda = \rho$
Brake → regen	$v = W$	$\lambda = \rho$

Differentiating Eq. (3.7) gives

$$\begin{aligned}\frac{d\lambda}{dx} &= \frac{1}{\hat{m}v}\frac{d\mu_2}{dx} - \frac{\mu_2}{\hat{m}v^2}\frac{dv}{dx} \\ &= \frac{\lambda\psi(v) - \psi(V)}{\hat{m}v^3}.\end{aligned}$$

In general, the value of λ determines the possible optimal driving modes. The driving mode determines how the speed of the train evolves, which in turn determines how λ evolves. Whenever λ reaches a critical value $\lambda = 1$ or $\lambda = \rho$, the control could change.

It is easier to understand the evolution of v and λ in the (λ, v) phase space. We will start by considering a simple journey on a straight, level track.

3.3.5 Optimal Journeys on a Straight, Level Track

Figure 3.4 shows a (λ, v) phase plot for an optimal optimal power—hold—coast—brake journey on a straight, level track with $G(x) = C(x) = 0$. The hold speed for the journey is $V = 30\,\text{ms}^{-1}$, and the regeneration efficiency is $\rho = 0.5$.

The journey starts to the right of the graph with $v = 0$ and large λ. The control mode is power, since $\lambda > 1$. As the speed of the train increases, λ decreases. If λ_0 is chosen correctly, the trajectory will follow the dark curve up and to the left towards the point $(1, V)$. There are three possible optimal transitions at this point:

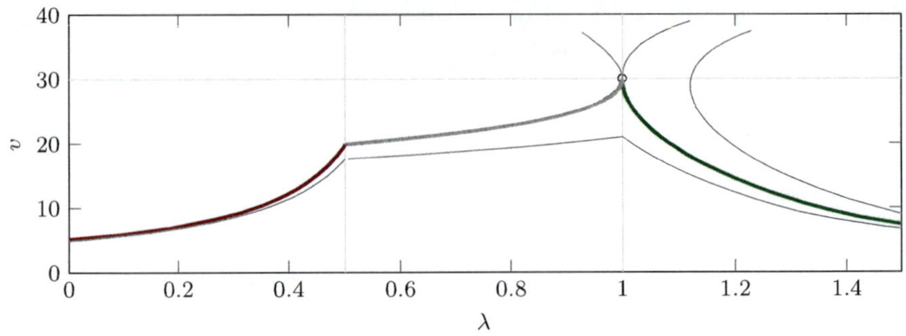

Fig. 3.4 Phase plot for an optimal power—hold—coast—brake journey on a straight, level track

- if the driving mode is kept as power then the speed of the train will continue to increase, λ will also increase, and there will be no more opportunities to switch out of power
- the driving mode can be changed to hold
- the driving mode can be changed to coast.

A hold phase can remain at $(1, v)$ indefinitely. There are no conditions that indicate when to switch out of a hold mode. However, the switching point out hold mode will determine where the train stops, since the following coast and brake phases are completely determined by switching point.

The hold phase could be omitted by switching straight from power to coast at $(1, V)$.

During a coast phase starting from $(1, V)$, both λ and v will decrease. There will be no more opportunities to switch back to hold or to power.

When the train state reaches $\lambda = \rho$ the control mode must change to brake. A regen mode is not possible on a level track because it is not possible to maintain the required constant speed with partial braking.

During the brake phase, both λ and v will continue to decrease. There will be no more opportunities to switch back to coast. Eventually, the train will stop.

The rightmost grey curve in Fig. 3.4 starts with λ_0 so high that λ never reaches $\lambda = 1$. In this case, the corresponding journey would remain in power forever.

The lowest grey curve in Fig. 3.4 starts with λ_0 low, so that the state reaches $\lambda = 1$ before the train speed reaches the $v = V$. In this case, the control must switch to coast before the train reaches the hold speed V.

Figure 3.5 shows the (x, v) profile for the optimal power—hold—coast—brake journey with $V = 30$.

If we want to complete this journey faster we can increase the hold speed V. Figure 3.6 shows the optimal profile for $V = 100$. This hold speed is higher than the train can achieve, so the train never reaches a state where it can switch to hold. The duration of the coast phase is quite short, and could be made even shorter by increasing V further.

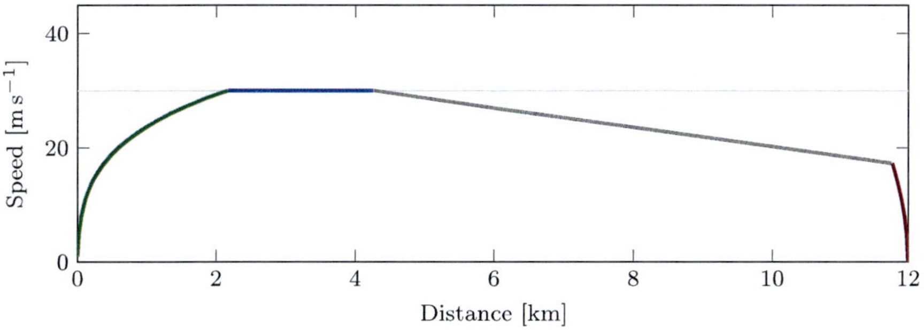

Fig. 3.5 Speed profile for an optimal power—hold—coast—brake journey on straight, level track

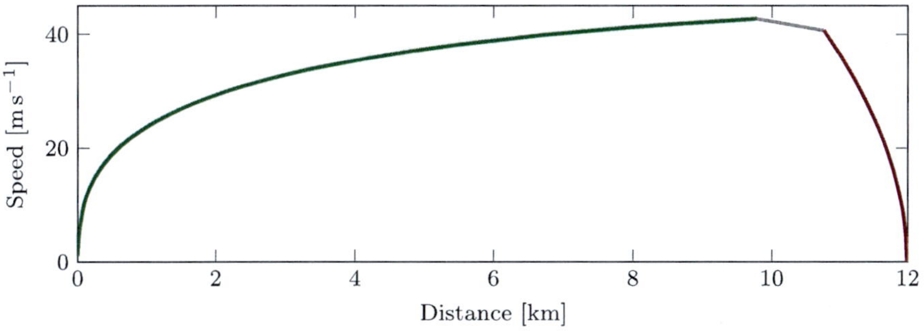

Fig. 3.6 Speed profile for an optimal power—coast—brake journey on straight, level track

Each possible hold speed V determines a unique journey that is optimal for some corresponding journey duration. The unique optimal journey that satisfies the desired journey duration can be found by adjusting V.

3.3.6 *Steep Inclines and Steep Declines*

Finding the optimal driving strategy is more difficult on tracks with steep gradients and speed limits. This section discusses the impact of steep gradients. The impact of speed limits is discussed in Sect. 3.3.7.

An incline is a *steep incline* at speed v if maximum tractive effort is not sufficient to maintain speed v. Similarly, a decline is a *steep decline* at speed v if the train speed increases while coasting on the decline.

Figure 3.7 shows an optimal hold—power—hold sequence for a steep 2% incline between 4 km and 6 km. The grey shaded region indicates the altitude profile of the track. The power phase starts before the start of the incline, and during the power phase speed increases before the incline, decreases on the incline, then increases again after the incline. The power phase starts at $v = V$ and $\lambda = 1$. The start location

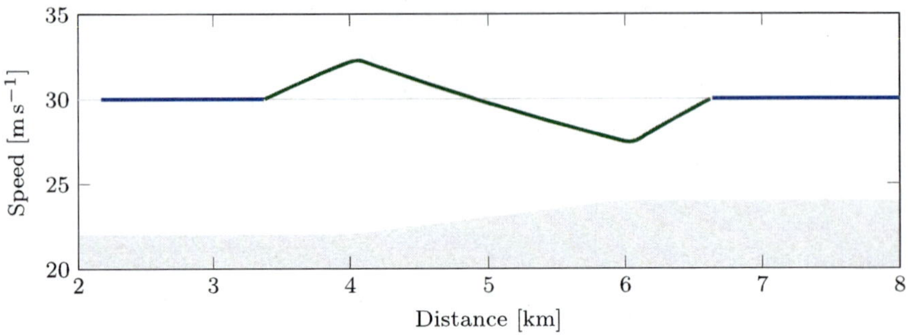

Fig. 3.7 Speed profile of an optimal hold—power—hold sequence for a steep incline

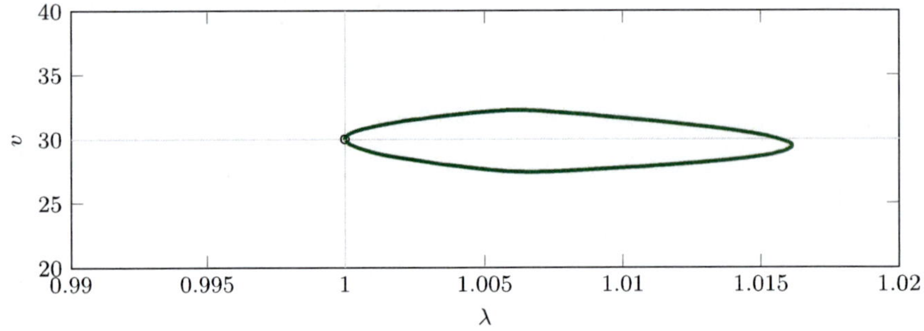

Fig. 3.8 Phase plot of the optimal hold—power—hold sequence for the steep incline

must be chosen so that the power phase also finishes with $v = V$ and $\lambda = 1$, so that the control can be switched back to hold. The optimal start location is found by searching. If the start location is too early then λ will not return to $\lambda = 1$, and if it is too late then $\lambda = 1$ will occur before speed v is back to $v = V$. Figure 3.8 shows the phase plot of the optimal power phase. The train state goes clockwise around the trajectory.

Figure 3.9 shows an optimal hold—coast—hold sequence for a steep -2% decline between 14 and 16 km. The coast phase starts before the start of the decline, and during the coast phase speed decreases before the decline, increases on the decline, then decreases again after the decline. The coast phase starts at $v = V$ and $\lambda = 1$. The start location must be chosen so that the coast phase also finishes with $v = V$ and $\lambda = 1$, so that the control can be switched back to hold. As before, the optimal start location is found by searching. Figure 3.10 shows the phase plot of the coast phase. The train state goes clockwise around the trajectory.

The previous two examples had isolated steep sections—the control intervals for the two steep sections did not overlap. The next example (Fig. 3.11) shows the optimal hold—power—coast—hold sequence for a 2% incline between 24 and 26 km, followed immediately by a -2% decline between 26 and 28 km.

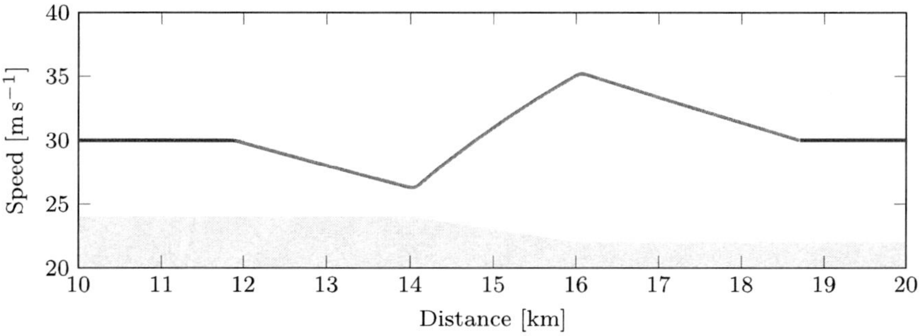

Fig. 3.9 Speed profile of an optimal hold—coast—hold sequence for a steep decline

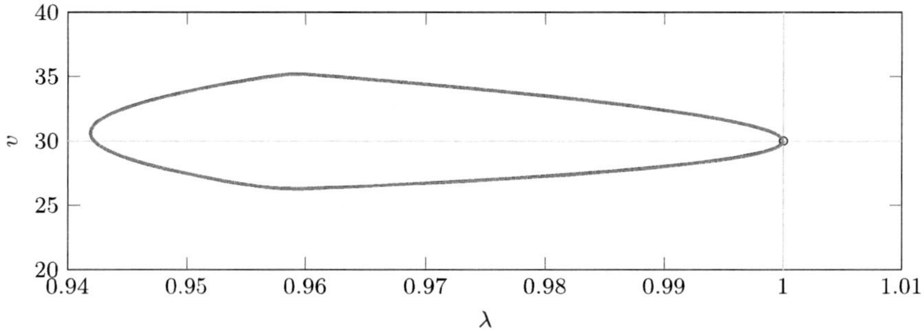

Fig. 3.10 Phase plot of the optimal hold—coast—hold sequence for the steep decline

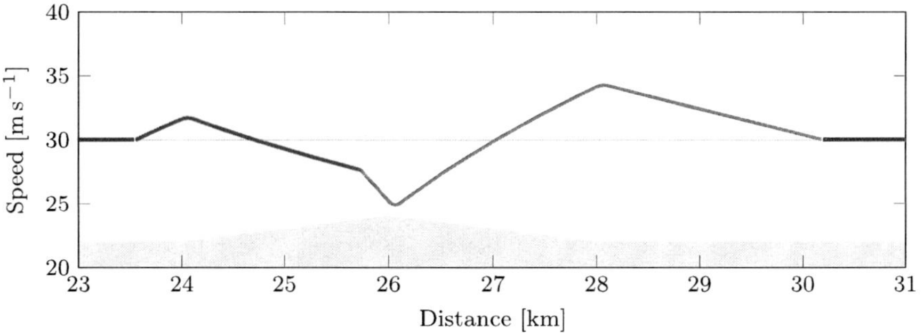

Fig. 3.11 Speed profile of an optimal hold—power—coast—hold sequence for a steep hill

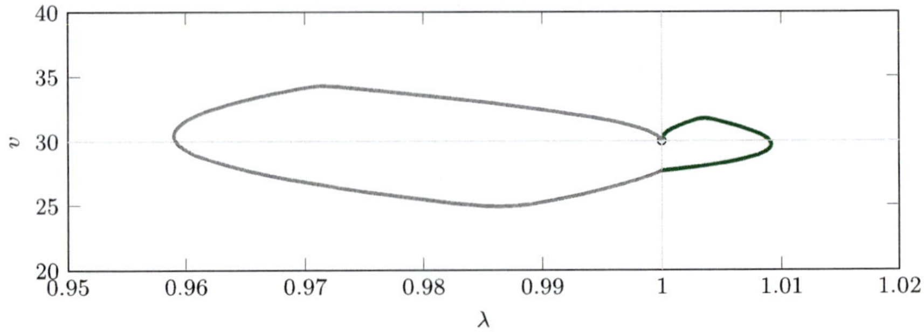

Fig. 3.12 Phase plot of the optimal hold—power—coast—hold sequence for the steep hill

The coast phase starts before the train reaches the crest of the hill at 26 km, so that the speed on the decline does not become too large.

Figure 3.12 shows the phase plot of the hold—power—coast—hold phase. The train state goes clockwise around the trajectory.

The general method for finding the optimal sequence of control modes and the optimal switching times for a journey with a predetermined hold speed V uses a numerical differential equation solver, such as a Runge-Kutta method, to calculate the evolution of speed v and adjoint λ, using the value of λ to determine the optimal control. The journey is calculated in stages separated by hold phases:

- The first stage starts with $v = v_0$ and λ_0 unknown. A search procedure is used to find the value of λ_0 that results in either:
  - the journey finishing with a brake phase that stops at the desired finish location, or
  - the journey reaching a state $(1, V)$ on a non-steep section of track so that the mode can change to hold.
- If the journey has reached a hold phase then the next stage starts with $v = V$ and $\lambda = 1$ but the end location of the hold phase unknown. A search procedure is used to find the starting location for the the next stage, which will finish either at the end of the journey or at another hold phase.

3.3.7 *Speed Limits*

Realistic journeys must adhere to speed limits, which may include:

- a train speed limit that applies throughout the journey because of speed limits of the locomotives or wagons
- permanent speed limits that apply to sections of the route because of bridges, curves, or track or substructure condition

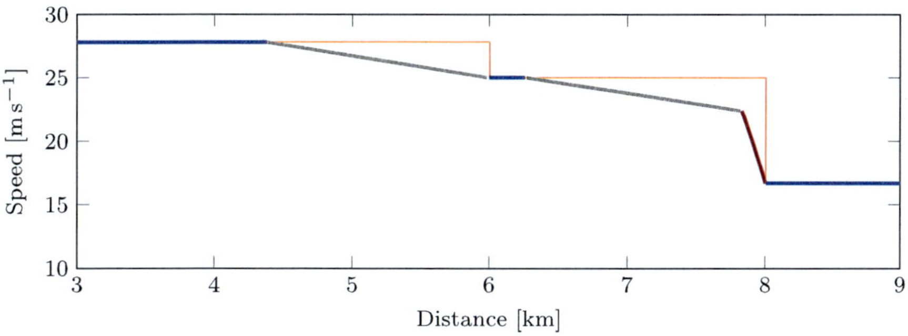

Fig. 3.13 Speed profile of an optimal control sequence with speed limits

- a speed limit that applies to the entire route because of weather conditions, such as high temperatures
- temporary speed restrictions that apply to sections of the route because of track condition or maintenance work.

Most speed limits apply to the entire length of the train—the train may not exceed a speed limit until the rear of the train has passed the end location of the limit. However, some speed limits, such as speed limits imposed to ensure adequate time to see and react to a track-side signal, apply only to the front of the train.

Just as with gradients, speed limits that apply to the entire length of the train can be transformed so that the train can be treated as a point mass train. For every location of the front of the train, the transformed speed limit is the lowest speed limit that applies at any location along the length of the train.

When a speed limit is encountered, the control must be adjusted so that the speed limit is not exceeded. The tricky part is deciding when to stop following a speed limit. This decision is similar to deciding when to depart from a hold phase [11].

Figure 3.13 shows an optimal speed limit—coast—speed limit—coast—brake—speed limit sequence when the speed limit drops from 100 km/h (27.8 m s^{-1}) down to 90 km/h (25 m s^{-1}) down to 60 km/h (16.7 m s^{-1}). Figure 3.14 shows the corresponding phase plot. The top curve corresponds to the first coast phase, and the bottom plot to the second coast—brake phase sequence.

Note that λ does not have to be continuous when a speed limit is encountered, so does not necessarily depart from the speed limit with $\lambda = 1$:

- on a steep decline, the train must either brake to a lower speed limit or else depart from the end of speed limit with λ found by searching
- on a steep incline, the train must depart from the beginning of the speed limit with λ found by searching
- on non-steep track, the train can depart from the speed limit by coasting before the end of the speed limit with $\lambda = 1$, or coast and brake with $\rho < \lambda < 1$ if required to avoid a future speed limit, or leave from the end of the speed limit with λ found by searching; only one of these will be possible.

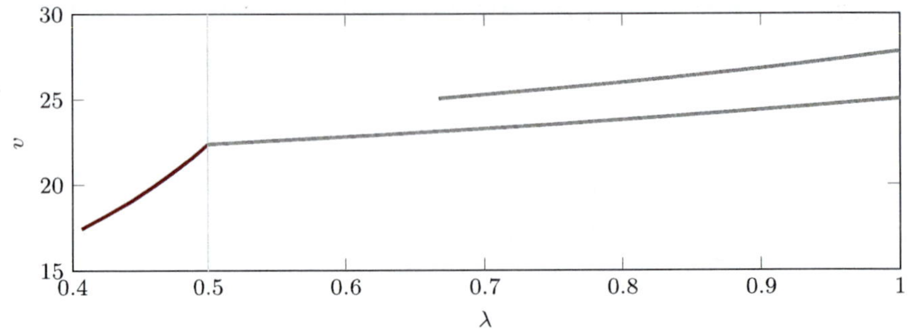

Fig. 3.14 Phase plot of the optimal control sequence with speed limits

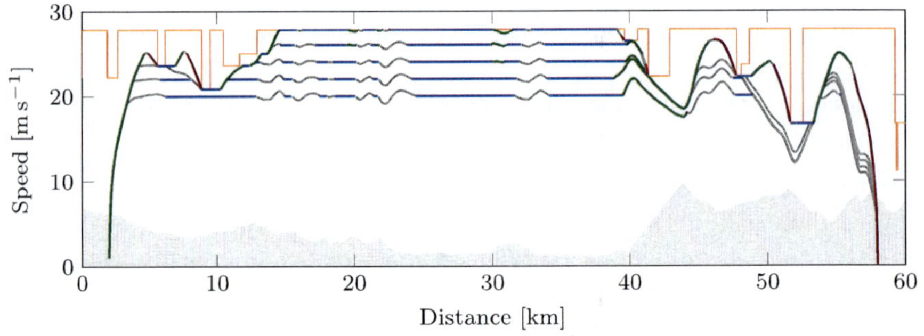

Fig. 3.15 Speed profiles for a freight journey with different hold speeds

3.4 Journey Duration and Energy

Each hold speed V gives a unique optimal journey, each with a unique journey duration. Figure 3.15 shows optimal speed profiles for a freight journey with hold speeds $V \in \{20, 22, 24, 26, 100\}$. The grey shaded region indicates the track altitude profile. The top, stepped curve is the track speed limit. The train comprises two 2700 kW diesel-electric locomotives hauling a 720 m, 2000 tonne load.

Increasing the hold speed V decreases the journey duration. Even when the hold speed V is above the maximum allowable track speed or the maximum attainable train speed, increasing V will decrease journey duration by decreasing the amount of coasting. As $V \uparrow \infty$, coasting durations approach zero and the total journey duration approaches a minimum.

Energy use for a journey increases as V increases and journey duration decreases. Figure 3.16 shows the relationship between journey duration and energy for the freight journey example as hold speed is varied from $V = 20$ to $V = 100$. This curve was found by calculating energy use for a set of optimal journeys with different durations. Howlett [8] derives a formula for the relationship between energy and journey duration for journeys without speed limits

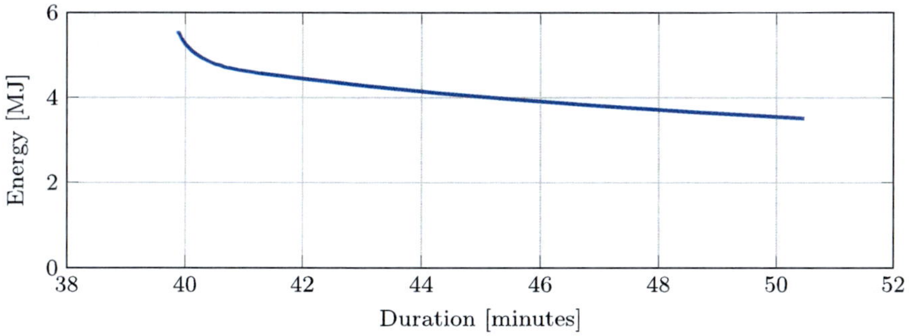

Fig. 3.16 Journey duration and energy for the freight journey with different hold speeds

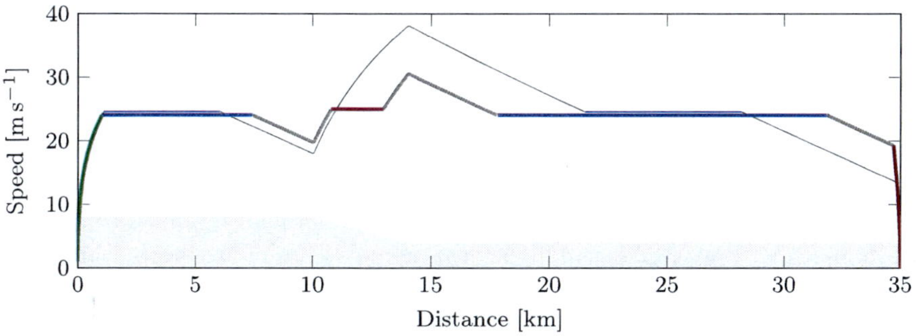

Fig. 3.17 Speed profiles for journeys with no regeneration (thin) and with 90% regeneration (thick)

3.5 Regeneration

The driving strategy for a train with regenerative braking is different to the strategy for a train without regenerative braking:

- coasting phases are shorter
- on steep declines, regeneration is used to limit the speed of the train.

Figure 3.17 shows two optimal speed profiles on a track with a 2% decline between 10 and 14 km. One train has no regenerative braking and the other has 90% regeneration. The journey duration is 25 min for both trains.

The train without regenerative braking has hold speed $V = 25.48$. The decline causes a long coast phase with large speed variation, and there is a long coast phase before braking at the end of the journey.

The train with 90% regeneration has a coast—regen—coast sequence for the decline, with smaller speed variation. It also has a shorter coast phase at the end of the journey. The hold speed is $V = 24.05$ and the regen speed is $W = 24.94$.

For the example in Fig. 3.17, the train with 90% regeneration efficiency following the regeneration strategy uses 190 MJ of energy for the journey, whereas the train

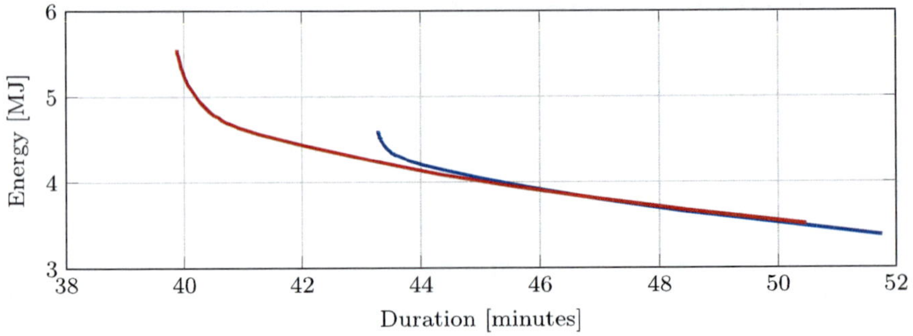

Fig. 3.18 Journey duration and energy for a freight journey with 2 locomotives (lowest duration 40 min) and with 1 locomotive (lowest duration 43 min)

without regeneration follows the no-regeneration strategy and uses 228 MJ of energy for the journey. If the train with regeneration were to follow the no-regeneration strategy it would still regenerate during the final braking phase and recover 21 MJ of energy, giving a total energy use of 207 MJ. In this case the regeneration strategy is better than the no-regeneration strategy for the train with regeneration because the train avoids high speeds and high aerodynamic losses on the steep decline. For journeys without steep declines the choice of driving strategy makes less difference to the overall energy use, and so it does not matter too much if the available regeneration efficiency is not know precisely.

3.6 Using More Power to Save Energy

Section 3.3 showed that the optimal control for a journey uses maximum tractive effort during power phases. This implies, perhaps counter-intuitively, that a train with more locomotives can use less energy for a journey [3].

Figure 3.18 shows journey duration and energy for two trains: the 720 m, 2000 tonne freight train hauled by 2 locomotives, as in Sect. 3.4, and the same trailing load hauled by 1 locomotive.

As expected, the train with 2 locomotives is able to achieve a lower journey duration than the train with 1 locomotive. For journey durations less than 47 min, the train with 2 locomotives uses less energy than the train with 1 locomotive.

Figure 3.19 shows the speed profiles for a 44-minute journey with 2 locomotives and with 1 locomotive. The train with 2 locomotives has a lower hold speed and smaller speed variations on steep gradients.

Train operating companies are sometimes tempted to require drivers to turn unnecessary locomotives offline or to avoid using high throttle notches on flatter sections of track. This is misguided. Even using more power available does increase the rate

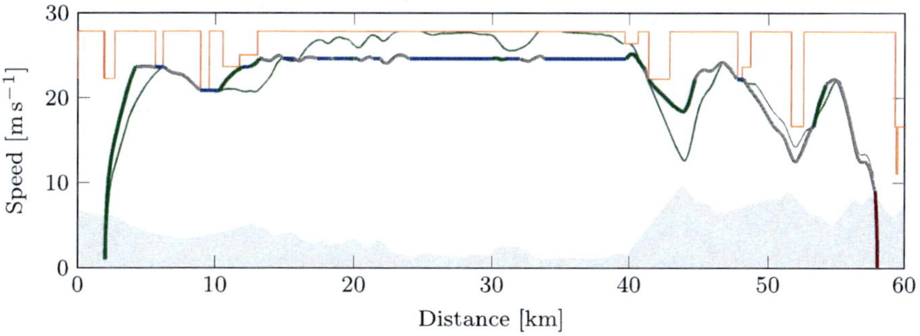

Fig. 3.19 Optimal speed profiles for a 44 min freight journey with 2 locomotives (thick) and 1 locomotive (thin)

Table 3.2 Duration and energy use of the initial power phase with 1 and 2 locomotives

Locomotives	Duration (s)	Energy (kJ)
1	285	568
2	150	599

of energy consumption, the time spent in power phases is reduced. Table 3.2 shows the duration and energy use of the initial power phase of the journeys in Fig. 3.19.

The train with 2 locomotives spends considerably less time getting up to speed. It uses slightly more energy during the initial power phase, but because of the time saved it can save more energy by travelling slower for the remainder of the journey. Overall energy use is 4.21 GJ with 1 locomotive and 4.13 GJ with 2 locomotives.

Even though the theory shows that having more power available is better, this applies only if the train is driven optimally. In practice, restricting power will limit the energy that can be wasted by driving inefficiently.

3.7 Intermediate Time Constraints and Timing Windows

Schedules for individual trains often contain timing targets for intermediate locations along a journey in addition to the desired arrival times at stops. The purpose of these intermediate timing targets may be simply to provide guidance to the driver, or could be necessary to maintain sufficient headway between trains, particularly at junctions.

To handle timing constraints, the journey is split into sections between adjacent timing locations. In general, a different hold speed V is required on each section. Finding the precise location at which to transition between two adjacent hold speeds, and hence the optimal speed of the train at the timing location, can be complicated. For example, a change to a lower hold speed will generally require a coasting phase

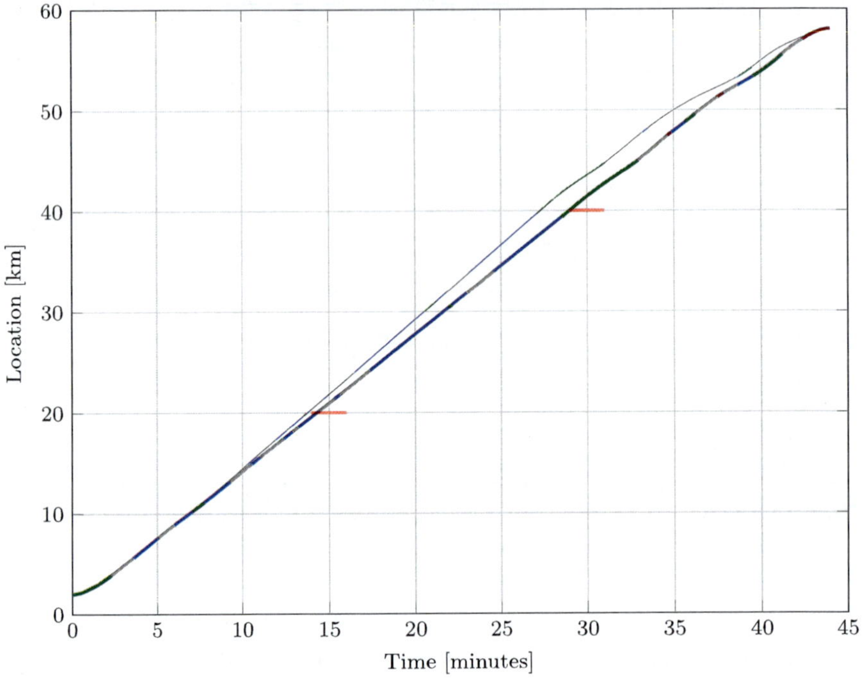

Fig. 3.20 Optimal non-stop freight journey without timing windows (thin) and with timing windows (thick). Timing windows are indicated by the short horizontal lines

that starts before the timing location and finishes after the timing location. In practice, changing the control at the timing location is usually accurate enough [12].

Albrecht et al. [1] describe time impact of timing windows on optimal train control and scheduling, and Albrecht et al. [2] describe how timing windows can be used to optimise gradual recovery from delays.

Figure 3.20 shows plots of time and location for an optimal journey without timing windows and for an optimal journey where the train is constrained to pass 20 km in the interval [14, 60] min and to pass 40 km in the interval [29, 31] min. Both journeys have the same overall journey time of 44 min. The journey with constraints has to slow down to meet the timing window at 40 km, and as a result also meets the timing window at 20 km. As expected, the journey without constraints uses less energy—4.13 GJ for the journey without constraints, and 4.43 GJ for the journey with timing windows.

The method of optimising through timing windows can also be used to find arrival times at stops that will minimise total energy for a journey, by allowing a wide arrival time window at each stop. The optimal journey has the same hold speed on every section. Train schedules are often developed with arrival times rounded to the nearest minute and with slack allocated to the last section, which leads to inefficient driving if the train is driven to meet this timetable. Figure 3.21 shows the difference in optimal

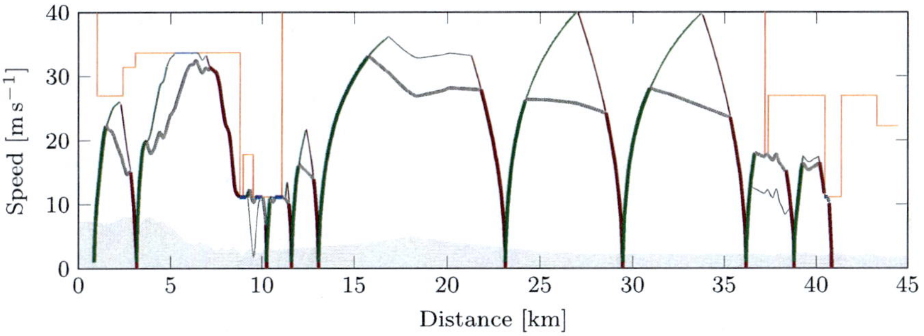

Fig. 3.21 Optimal journey for a passenger train using the original train schedule (thin) and an optimised schedule (thick)

speed profiles for a passenger train driven to the original timetable and to an optimised schedule. The optimised schedule gives lower variation in speeds between sections. Energy for the journey is 1.73 GJ for the original train schedule and 1.05 GJ with the optimised arrival times. Energy-efficient train timetabling is covered in more detail in the next chapter.

3.8 Driving Advice Systems

The rules for constructing an optimal journey are complicated, and calculating an optimal journey profile requires extensive use of numerical shooting methods to find trajectories that meet the necessary conditions for an optimal journey. Despite these difficulties, calculations based on these specialised methods are much faster than general search methods and guaranteed to find an optimal solution. Optimal journeys spanning hundreds of kilometres, with many gradient and speed limit changes and many timing constraints, can be calculated in seconds [5, 14]. This means that these calculation methods can be used to calculate driving advice in real time; they have been used in the TTG Energymiser driving advice system since 2008.

The data required by a driving advice system includes:

- train data: tractive effort curves, braking effort curves, train length and mass, and resistance coefficients
- route data: elevation profile, curves, speed limits, and geodetic coordinates used to convert GPS position to location along the route
- schedule data: departure location and time, arrival time windows at timing locations and station dwell times.

This information is typically prepared on a central server and uploaded to trains prior to the start of a journey. During the journey, GPS position information is used to determine the location and speed of the train, typically at 1-second intervals. Optimal driving profiles are calculated by the on-board system from the current time, location and speed to the next stop. Driving advice can be displayed to the driver, or sent to an automatic train operation (ATO) system. Energymiser continuously compares the state of the train to the ideal profile; if the train is more than 10 s ahead or behind schedule or if the speed of the train is more than a few km/h above or below the ideal speed then a new optimal profile is calculated within seconds.

Driving advice systems have been deployed by many rail operators on a wide variety of rail systems, including long-haul mixed freight, bulk freight, suburban and intercity passenger trains, and very fast trains. Energy savings are highly dependent on the type of operation, on the amount of slack in the schedules and on driver engagement, but typically lie in the range 5–20%.

While the original aim of driving advice systems was to help drivers reduce energy use, the ability to keep trains on time is sometimes even more important. It is crucial that driving advice systems are given feasible and efficient train schedules to work to, and the methods described in this chapter and in following chapters can be used to allocate slack to a train schedule in a way that will ensure efficient driving and on-time arrivals.

The next phase for driving advice systems is connected driving advice systems (C-DAS). These systems incorporate real-time communication between on-train DAS and central traffic management systems to allow real-time updates to train schedules. Two applications of C-DAS currently being developed are:

- junction scheduling: trains approaching a junction are monitored, and target arrival times at the junction are adjusted in real time to ensure sufficient headway between trains to avoid restrictive signals [7]
- peak demand management: all of the electric trains on a network can have their energy use and progress monitored and their schedules adjusted in real time to reduce the network's demand for electricity during critical intervals for the electricity grid.

3.9 Conclusion

Being able to calculate energy-efficient driving strategies has important applications. First, it enables the calculation of journey schedules that distribute slack in a way that ensures that each train can be driven at a consistent pace throughout its journey, which will improve the efficiency and reliability of journeys. Second, it can be used to provide real-time driving advice to ensure that trains are driven precisely to the planned timetable, which will reduce delays, increase service reliability and increase network capacity. Finally, it can give significant reductions in energy use and the associated greenhouse gas emissions.

References

1. Albrecht T, Binder A, Gassel C (2013) Applications of real-time speed control in rail-bound public transportation systems. IET Intell Transp Syst 7(3):305–314
2. Albrecht A, Howlett P, Pudney P et al (2013) Using timing windows to allow energy-efficient recovery from delays. In: 10th world congress on rail research
3. Albrecht A, Howlett P, Pudney P et al (2014) Using maximum power to save energy. In: 17th international IEEE conference on intelligent transportation systems (ITSC). IEEE, pp 1205–1208
4. Albrecht A, Howlett P, Pudney P et al (2016) The key principles of optimal train control-part 1: formulation of the model, strategies of optimal type, evolutionary lines, location of optimal switching points. Transp Res Part B Methodol 94:482–508
5. Albrecht A, Howlett P, Pudney P et al (2016) The key principles of optimal train control-part 2: existence of an optimal strategy, the local energy minimization principle, uniqueness, computational techniques. Transp Res Part B Methodol 94:509–538
6. Davis WJ (1926) The tractive resistance of electric locomotives and cars. Gen Electr
7. Galapitage A, Albrecht AR, Pudney P et al (2018) Optimal real-time junction scheduling for trains with connected driver advice systems. J Rail Transp Plann Manage 8(1):29–41
8. Howlett P (2016) A new look at the rate of change of energy consumption with respect to journey time on an optimal train journey. Transp Res Part B Methodol 94:387–408
9. Lukaszewicz P (2001) Energy consumption and running time for trains: modelling of running resistance and driver behaviour based on full scale testing. KTH
10. Milroy IP (1980) Aspects of automatic train control. Loughborough University
11. Pudney P, Howlett P (1994) Optimal driving strategies for a train journey with speed limits. ANZIAM J 36(1):38–49
12. Pudney P, Howlett PG, Albrecht A et al (2011) Optimal driving strategies with intermediate timing points. International Heavy Haul Association
13. Scheepmaker GM, Goverde RMP, Kroon LG (2017) Review of energy-efficient train control and timetabling. Euro J Oper Res 257(2):355–376
14. Scheepmaker GM, Pudney PJ, Albrecht AR et al (2020) Optimal running time supplement distribution in train schedules for energy-efficient train control. J Rail Transp Plann Manage 14:100180

Chapter 4
Energy-Efficient Train Timetabling

Rob M. P. Goverde and Gerben M. Scheepmaker

4.1 Introduction

The aim of energy-efficient train timetabling is to construct a timetable that enables energy-efficient train operation. The train timetabling problem determines for each train the arrival and departure times at all stations or other timing points, while considering track capacity constraints such that the resulting train paths are conflict-free [7, 21]. The energy consumption of the trains depends on the allocation of the time allowances over the timetable. First, the distribution of running time supplement over a train path affects the train driving behaviour and therefore the energy consumption. Second, the allocation of buffer times between the scheduled train paths determines the robustness of a timetable, such that small delays do not immediately result in path conflicts. This avoids an increase in energy consumption due to braking or unplanned stops followed by re-acceleration and running faster to recover the time loss.

In the train timetabling problem the train paths for all train services are scheduled and coordinated aiming at a conflict-free allocation of trains to the railway capacity. A train path is the time-distance allocation of a train service to a specific sequence of tracks and station platforms over the railway infrastructure, including the arrival and departure times at the stations, as well as possible additional passage times at intermediate timing points. The majority of the literature on train timetabling is about computing feasible and robust timetables for given lower and upper bounds on the activity times, such as dwell times at stations, running times between stations, transfer times between connecting trains at stations, and minimum headway times between event times of adjacent train paths [7, 21]. The running times are usually

R. M. P. Goverde (✉)
Department of Transport and Planning, Delft University of Technology, Delft, Netherlands
e-mail: R.M.P.Goverde@tudelft.nl

G. M. Scheepmaker
Netherlands Railways, Utrecht, Netherlands
e-mail: gerben.scheepmaker@ns.nl

S. Su et al., *Energy-Efficient Train Operation*, Lecture Notes in Mobility,
https://doi.org/10.1007/978-3-031-34656-9_4

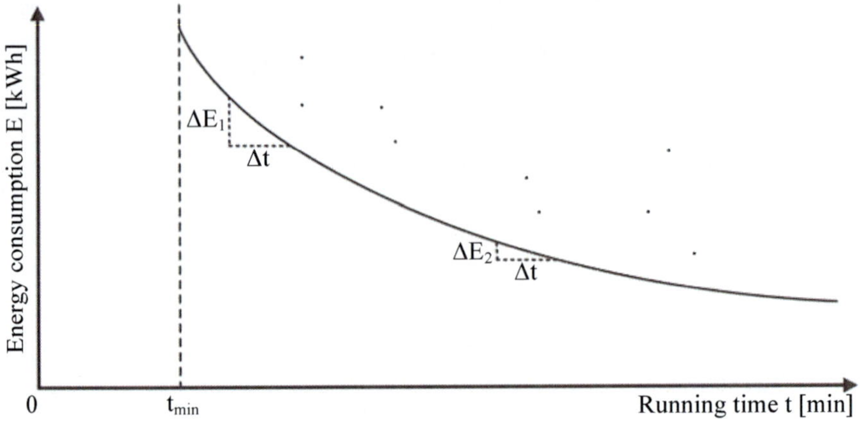

Fig. 4.1 Pareto front of energy consumption versus running time

fixed or given some flexibility, based on an external running time calculation for the minimum technical running time and norms for the running time supplement. Optimization of the nominal timetable usually aims at minimizing the travel times (including dwell and transfer times), while the robust train timetabling problem aims at reducing delay propagation by including sufficient buffer times between the train paths. The optimization of both the nominal and robust train timetabling problem thus focuses on the time domain, while the impact of changes in running times to the speed profiles and therefore to energy consumption, either negative or positive, is mostly discarded. Likewise, the impact of a change in running time to the minimum headway times is also hardly considered in the existing literature, which would also require a microscopic level of detail similar to train trajectory optimization.

The amount of scheduled running time determines the running time supplement available for energy-efficient driving. The minimum running time corresponds to running as fast as possible considering the train and track characteristics and respecting all operational constraints, which also leads to the highest energy consumption. Running time supplement is added to increase the stability of the timetable by (1) reducing the occurrence of primary delays due to train parameter variations and external (weather) conditions by which a train needs more time than scheduled, and (2) recovering existing delays by running faster than scheduled [13]. The added running time supplement must be translated into a 'slower' speed profile that covers the scheduled running time and cumulatively determines the running time over the entire train path. The exact speed profile determines the resulting energy consumption.

Figure 4.1 shows a typical Pareto front corresponding to the trade-off between energy consumption and running time. The energy consumption is the highest for the minimum running time $t_{\min}$. For larger running times the energy consumption depends on the driving behaviour. The Pareto front corresponds to the energy-efficient speed profiles, which dominate other solutions associated to non-optimal driving (indicated by the dots). The decreasing Pareto curve illustrates that increasing the running time leads to energy savings, but this effect decreases for larger running times, i.e., increasing the running time by some amount Δt will lead to a larger energy saving ΔE_1 than the energy saving ΔE_2 obtained by a further increase by the same amount when already supplement is included, $\Delta E_1 > \Delta E_2$. These Pareto curves can be used to find the optimal allocation of running time supplements over multiple stops, by first generating the Pareto fronts for all train runs between two stops (or other timing points) and then selecting the running times from the Pareto curves to the train runs such that the total energy is minimized [3, 9, 11, 31–33, 38]. The first step can be done by solving the energy-efficient train trajectory optimization problem between two stops for various running times for each train path segment, while the second step can be done using dynamic programming [3] or direct search algorithms.

A train trajectory optimization model can also be used to compute the optimal energy-efficient speed profile over multiple stops directly, including the optimal arrival and departure times at the intermediate stops [6, 30, 35, 36]. When multiple trains are considered track capacity constraints must be modelled, which can take the form of default headway times at arrivals and departures as in macroscopic train timetabling problems [19, 35, 36, 41] or dynamic minimum headway times computed after the train trajectories have been computed [28, 40]. An alternative optimization model integrates the train timetabling and speed profile problem in a space-time-speed network, where the speed profiles are linearized or otherwise simplified to keep the resulting optimization problem tractable [40, 43]. As dwell (and transfer) times are a main source of disturbances or delays affecting the remaining running time supplement for energy-efficient driving, also models have been proposed to include stochastic or fuzzy dwell times in the train trajectory optimization problem over multiple stops [6, 9, 15], while other models include passenger flows over the network [19, 39, 41]. For an extensive literature review of energy-efficient train timetabling, see Scheepmaker et al. [29].

In general, energy-efficient train timetabling leads to multi-objective optimization problems. Goverde et al. [15] developed a sequential approach for the multiple-objective timetable optimization problem on a network, where the first aim is to determine a conflict-free, stable and robust timetable. A micro-macro iterative model was developed to compute an optimal network timetable, where the macroscopic level optimizes a trade-off between travel times and robustness, and the microscopic level guarantees feasibility and stability using blocking time theory [22]. In a third step, energy-efficient speed profiles are embedded over the corridors between main stations, maintaining the scheduled departure and arrival times at the corridor ends and including stochastic dwell time distributions at intermediate stops. Scheepmaker and Goverde [28] proposed a multi-objective optimization problem for computing an energy-efficient train timetable on a heterogeneous traffic corridor considering the joint objectives of total running time, infrastructure occupation, robustness, and energy consumption. They first computed the energy-efficient train trajectories for a range of scheduled running times for each train type, followed by computing the associated microscopic blocking times. The latter were used to compute the minimum line headway times to guarantee conflict-free train paths and the infrastructure occupation to guarantee sufficient buffer time for robustness. The problem was then solved using the weighted-sum method.

Modern electric trains can apply regenerative braking using the engine as generator to convert kinetic energy into electricity. This regenerated energy can be used within the train (for lighting, heating, ventilation and air conditioning), stored in batteries, or fed back to the power supply system to be used by nearby trains [12]. In general, however, power is lost due to generated heat and the efficiencies of the engine, converters, energy storage device and/or the power supply system, depending on specific conditions. In this chapter, we exclude energy gains by regenerative braking, while similar results are obtained when regenerative braking is included in the objective function, see Scheepmaker and Goverde [27]. The main uncertainty is the determination of the regenerative braking efficiency, which requires detailed modelling of the energy storage system (see Chap. 6) or simulation of the power supply network (see Chap. 7). When regenerative braking is used, synchronization of arrival and departure times can increase the efficiency of reused regenerative braking energy by nearby accelerating trains, which will be considered in Chap. 5. In the present chapter, we focus on facilitating energy-efficient driving of each train separately without dynamic coordination or cooperation with other trains.

This chapter focuses on the modelling of train trajectory optimization problems for energy-efficient train timetabling. Algorithms for solving the optimal control problems based on the application of Pontryagin's Maximum Principle [1, 2, 16] are presented in Chap. 3. The examples in this chapter were all solved by a pseudospectral method using Pontryagin's Maximum Principle for validation [16, 34].

The structure of this chapter is as follows. Section 4.2 first considers the minimum-time train trajectory problem which is used as a reference driving strategy and applied in practice when a train is delayed. Section 4.3 considers the energy-efficient train trajectory optimization problem between two stops. This is extended in Sect. 4.4 to energy-efficient train trajectory optimization over multiple stops, where the running times between successive stops are optimized. Section 4.5 then continues with energy-efficient train timetabling over corridors with multiple trains. Finally, conclusions on energy-efficient train timetabling are presented in Sect. 4.6.

4.2 Minimum Running Time Calculation

4.2.1 Problem Formulation

The minimum running time is the technical running time that a train needs to traverse a given route between two timing points as fast as possible, while considering the basic characteristics of the train and track. It assumes default values for the train and track parameters representing good conditions, such as good weather conditions and constant power supply. The minimum running time is the basis for determining the scheduled running times in the timetable and it is assumed to be achievable when a train is running late. Therefore, the minimum running time is essential information for both timetable planning and operational traffic management. This section will focus on the minimum running time between two station stops. Minimum running times between arbitrary timing points, such as signals, junctions or non-served stations, are obtained in a similar way with a positive speed specified instead of standstill at a station platform track.

The minimum running time calculation is based on data about the train, track and operations:

- Train characteristics: mass, length, maximum speed, resistance, traction, brake,
- Route-specific track characteristics: speed limits, gradients, curves, tunnels,
- Operational characteristics: planned stops, signalling.

A train consists of a formation of one or more coupled rolling stock units, which may be either self-propelled multiple units or locomotives with carriages. The train characteristics therefore depend on a specific train composition, such as train length and mass. The traction, or tractive effort, $F(v)$ [N] is the sum of all tractive forces at the driving wheels and is speed-dependent. It is usually given as a nonlinear tractive effort curve in a force-speed diagram that represents the maximum tractive effort $F^{\max}(v)$ as a function of speed v [m/s]. The curve typically includes a constant part at low speeds corresponding to the adhesion limit, and a hyperbolic part corresponding to the maximum power P [W] at higher speeds due to the relationship $F^{\max} = P/v$ until the maximum speed $v^{\max}$. The train resistance R [N] consists of rolling and air resistance which together form a a second-order polynomial of speed $R(v) = a + bv + cv^2$, with with non-negative coefficients $a, b \geq 0$ and $c > 0$, known as the Davis equation [10]. The mechanical braking characteristics are often given by a constant braking rate or a step function of braking rates as function of speed $B(v)$ [N] that models the maximum (service) braking. Also detailed formulae exist with parameters including a braking percentage and brake build-up time. In addition to mechanical braking, regenerative braking can be considered using the engine as generator, which can be modelled as a nonlinear function of speed like the opposite of the maximum tractive effort curve. Without loss of generality, we exclude regenerative braking in this chapter, but similar results are obtained when regenerative braking is included in the objective function [27]. Certain assumptions have to be made for the computation of the minimum running time, such as train mass for a typical passenger load, train resistance parameters for good weather conditions, and tractive forces for a given constant power supply.

An accurate running time calculation requires input about the planned route with the associated track description including the static speed profile, gradient profile, curve profile, tunnel sections, and possible other special sections such as power supply changes in electric railways. The gradients, curves and tunnels add additional line resistances to the train movement. The gradient profile is usually assumed to be a step function of distance corresponding to successive track sections with constant gradient. The gradient resistance for a slope with angle α is $G(s) = gm \sin \alpha \approx gmn$ [N], with $g = 9.81$ m/s^2 the acceleration due to gravity, m the train mass, and n [m/km] the change in vertical height with horizontal distance.

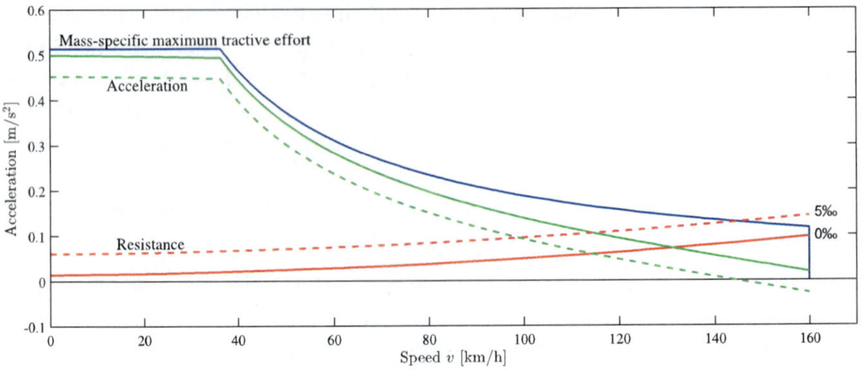

Fig. 4.2 Acceleration-speed diagram of mass-specific maximum tractive effort (blue), resistance and resulting acceleration for flat track (solid red and green) and $5^0\!/_{00}$ gradient (dashed red and green)

The force available for acceleration, known as the surplus force or acceleration force, is given by the force equilibrium of tractive effort minus total resistance (including line resistance of slopes, curves and tunnels) and includes a dimensionless rotating mass factor ρ,

$$\rho ma = F(s) - R(v) - G(s).$$

Instead of forces, it is convenient to consider the mass-specific forces that are obtained by dividing the forces by a factor ρm, which results in acceleration units m/s^2, and

$$a = (F(v) - R(v) - G(s))/(\rho m) = f(v) - r(v) - g(s).$$

Figure 4.2 shows an acceleration-speed diagram illustrating typical curves for mass-specific maximum tractive effort $f^{\max}(v)$ and resistance $r(v) - g(s)$, and the resulting mass-specific surplus force or acceleration $a(v)$. It clearly shows that acceleration is a nonlinear function of speed that decreases towards zero for larger speeds. The maximum braking rate for this train type is -0.66 m/s^2, illustrating that train acceleration is even much slower than braking. The acceleration curve on an incline with $5^0\!/_{00}$ shows that the maximum speed can even be limited on uphill slopes, although the maximum speed of this train type is limited to 140 km/h.

The static speed profile is a step function of distance that includes line speed limits and possible route-dependent speed restrictions due to reverse switches leading to a different track. Locations on the route affecting the operational speed must also be specified, such as the stop position of the train front at platform tracks (that may depend on the train length or number of carriages), and possible trackside signs or signals with a brake indication such as at the entry of stations. Moreover, the actual train behaviour depends on operational rules, such as starting with full acceleration only after the entire train has passed a reverse switch section and braking according to the requirements of the automatic train protection. For instance, with modern distance-to-go braking curve supervision, such as in the European Train Control System (ETCS), braking can start at the last moment before a location with a restricted (or zero) speed as long as the permitted braking curve is not exceeded. In contrast, in a traditional three-aspect fixed-block signalling system a train may need to start braking when passing a yellow approach signal and proceed with a restricted speed until a final braking regime to reach standstill at the stop position.

The minimum running time calculation can now be formulated as an optimal control problem. The minimum time train control (MTTC) problem from initial position s_0 to final position s_1 is then formulated as follows:

$$\text{Minimize } t(s_1) \tag{4.1}$$

subject to the constraints

$$\dot{t}(s) = 1/v(s) \tag{4.2}$$

$$\dot{v}(s) = (u(s) - r(v) - g(s))/v(s) \tag{4.3}$$

$$0 \leq v(s) \leq v^{\max}(s) \tag{4.4}$$

$$u^{\min}(v) \leq u(s) \leq u^{\max}(v) \tag{4.5}$$

$$t(s_0) = 0,\ v(s_0) = 0,\ v(s_1) = 0, \tag{4.6}$$

while the final time $t(s_1)$ is free. The independent variable is distance s [m], the state variables are time t [s] and speed v [m/s], $\dot{t} = dt/ds$ and $\dot{v} = dv/ds$ denote the derivatives of the state variables with respect to the independent variable s, and the control variable u [m/s^2] is the mass-specific applied (tractive or braking) force $u(s) = F(s)/(\rho m)$, i.e., the applied force divided by total mass including a rotating-mass factor ρ. The control is bounded between a maximum braking rate $u^{\min} = B/(\rho m) < 0$ and a maximum specific traction force $u^{\max} = F^{\max}/(\rho m) = \min(u_0, p^{\max}/v)) \geq 0$, with mass-specific adhesion limit u_0 and mass-specific maximum traction power $p^{\max} = P/(\rho m)$ [m^2/s^3] using the relation $p^{\max} = u^{\max} v$. Note that traction and braking cannot be used at the same time. We use the notation $u^+(s) = \max(u(s), 0) \geq 0$ and $u^-(s) = \min(u(s), 0) \leq 0$ so that $u(s) = u^+(s) + u^-(s)$. The resistance forces consist of a mass-specific train resistance $r(v) = R(v)/(\rho m)$ [m/s^2] and a mass-specific line resistance $g(s) = G(s)/(\rho m)$ [m/s^2]. Finally, the speed is bounded above by a speed limit $v^{\max}(s)$, which is assumed piecewise constant.

An alternative model could be provided with time as independent variable, and distance and speed as time-dependent state variables, which would lead to more conventional differential equations to time. However, it is more convenient to use distance as independent variable according to the track-oriented constraints such as the static speed and gradient profiles that are stepwise functions of distance.

4.2.2 Optimality Conditions

According to Pontryagin's Maximum Principle [24], the optimal control should maximize the Hamiltonian H defined as

$$H(t, v, \lambda_1, \lambda_2, u, s) = \lambda_1/v + \lambda_2\,(u - r(v) - g(s))\,/v, \tag{4.7}$$

where λ_1 and λ_2 are the co-state variables corresponding to time and speed, respectively. In the case of free final time, $\lambda_1 \equiv -1$, and λ_2 is a nonlinear function of the independent variable s [16]. This Hamiltonian is linear in the control u with coefficient λ_2/v. The optimal control structure $\hat{u}$ is therefore obtained from the sign of λ_2 as (1) maximum traction $u^{\max}(v(s))$ if $\lambda_2(s) > 0$, (2) cruising at the maximum speed using partial traction $r(v^{\max}(s)) + g(s)$ if $\lambda_2(s) = 0$, and (3) maximum braking $u^{\min}(v(s))$ if $\lambda_2(s) < 0$. This leads to the following optimal control structure [16]:

$$\hat{u}(s) = \begin{cases} u^{\max}(v(s)) & \text{if } \lambda_2(s) > 0 \text{ (Maximum acceleration)} \\ r(v^{\max}(s)) + g(s) & \text{if } \lambda_2(s) = 0 \text{ (Cruising at maximum speed)} \\ u^{\min}(v(s)) & \text{if } \lambda_2(s) < 0 \text{ (Maximum braking).} \end{cases} \tag{4.8}$$

The switching points between the three driving regimes are determined by the dynamics of the co-state variable λ_2 that satisfies an adjoint differential equation [16]. However, from the optimal control structure (4.8) a constructive method can be derived without the need to calculate the co-state λ_2 explicitly, which defines the fastest running: accelerate as fast as possible until the maximum speed is obtained, maintain the maximum speed as long as possible, and brake as fast as possible at the end to a standstill at the final destination s_1. Cruising at the maximum speed generally implies applying partial traction to counter the resistances at the maximum speed. On steep downhill slopes, where the speed increases even when not applying any traction, partial braking must be applied to maintain at the maximum speed. For safety reasons, the negative gradients should be limited such that maximum braking should prevent exceeding the maximum speed. If the static speed profile includes one or more intermediate speed restrictions then the train should brake as late as possible before the speed restriction and re-accelerate as early as possible again to a higher speed when possible. In case of steep uphill slopes where even full traction cannot keep the maximum cruising speed, the optimal control switches to maximum acceleration until the maximum speed is reached again (after the gradient has become less steep). In case of any restrictions from the signalling system the maximal permitted speed curves should be followed. Hence, this optimal control structure is completely aligned with the common technical running time calculation.

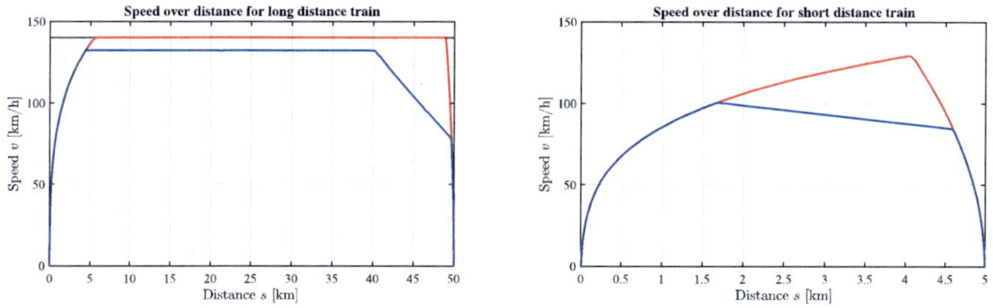

Fig. 4.3 Speed-distance diagrams for the minimum-time (red) and energy-efficient train control (blue) of a long-distance train (left) and a short-distance train (right)

4.2.3 Illustrative Example

Figure 4.3 shows speed-distance diagrams for typical minimum-time train trajectories for a long-distance train (left) and a short-distance train (right) on flat track. Also shown are the corresponding energy-efficient trajectories, which will be explained in the next section.

The minimum-time speed profile on the left shows a train accelerating to the speed limit $v^{\max} = 140$ km/h with maximal tractive effort bounded by the maximum traction-speed curve, then cruising at the speed limit using partial traction to counter the train and line resistances, and finally braking at the maximum braking rate to a standstill at the final position. This speed profile is similar for routes with non-steep gradients. Only the energy consumption would change due to varying partial traction in the cruising regime as opposed to a constant partial traction setting on flat track.

The speed-distance diagram on the right in Fig. 4.3 illustrates the minimum-time speed profile over a short stop distance. In this case, the distance is too short to reach the speed limit. The train now applies maximum acceleration until it hits the braking curve at which the control switches to maximum braking in order to reach standstill at the stop.

4.3 Energy-Efficient Train Trajectory Optimization Between Stops

4.3.1 Problem Formulation

In practice, scheduled running times consist of the minimum running time plus a running time supplement. The running time supplement serves various purposes:

1. To cover variations in the train parameters that will lead to larger running times,
2. To cover less favorable conditions such as strong headwind,
3. To compensate for variations in driver behavior,

4. To round up to full minutes (or any other time unit),
5. To fit the train path in the timetable at bottlenecks,
6. To reduce traction energy consumption,
7. To enable delay recovery.

The first three points relate to reliability of the scheduled running time. If the minimum running time would be used in timetables then delays will occur often as individual rolling stock will have (slightly) varying characteristics due to different states of wear. Furthermore, operational conditions may vary with fluctuating passenger flows and weather conditions, and also the driver behavior varies from day to day and between different drivers. Also the actual train composition may differ from the one used in the calculation of the minimum running time, which specifically holds for periodic timetables where the same scheduled running time is used for all trains from a specific train line throughout a day, while in practice the train length (and hence composition) is varied to accommodate fluctuating passenger demand over the day. In principle, the minimum running time could be computed for the worst-case train composition and conditions, but this will also lead to implicit running time supplements for those trains running with better train compositions and conditions. Therefore, the minimum running time is usually computed based on average parameter values and then a regular running time supplement is defined as a percentage of the computed minimum running time. Typical regular running time supplements are 5%–10% depending on specific rules from different railways or countries, which may differentiate for instance between short and long distance trains. In addition to this percentage, an additional absolute time supplement may be added for rounding or 'bending' the train path to fit between other train paths according to the 4th and 5th mentioned points.

Given a scheduled running time including running time supplement, the speed profile should be adapted for on-time running. This gives opportunities for energy-efficient driving which is thus considered as a secondary objective. In addition, if a train is running late then it could run faster to recover the delay by the actual running time supplement, which further improves punctuality and service stability. However, note that worst-case trains and conditions might actually need the full running time supplement, and thus do not have this recovery opportunity. Therefore, the actual combination of relative and absolute running time supplement is a design choice to enable reliable train services regarding both parameter variations and delay recovery. In particular, extra supplements may be planned just before bottlenecks or main stations to allow improved on-time running at these strategic locations. However, it then becomes very important that trains do not arrive early at these locations and disturb other trains, so that computing a feasible speed profile covering the entire scheduled running time under normal conditions remains essential.

The actual running time supplement may also be affected by a (slight) initial departure delay, which essentially reduces the actual available running time. Hence, optimal operational speed profiles may be computed depending on specific train characteristics and the available running time consisting of the scheduled running time minus any initial departure delay.

The energy-efficient train control problem or train trajectory optimization problem is the problem of finding the optimal train trajectory for a train run between two stops in a given scheduled time T [s] such that the total traction energy is minimized. The traction energy can be computed as the integral of the applied traction over distance. Hence, the objective function is

$$\text{Minimize} \int_{s_0}^{s_1} u^+(s)ds \tag{4.9}$$

under the constraints (4.2)–(4.5) and the boundary conditions

$$t(s_0) = 0, t(s_1) = T, v(s_0) = 0, v(s_1) = 0. \tag{4.10}$$

Note that in contrast to (4.6) the final time is given.

4.3.2 Optimality Conditions

According to Pontryagin's Maximum Principle [24] the optimal control should maximize the Hamiltonian H which is defined for this problem as

$$\begin{aligned} H(t, v, \lambda_1, \lambda_2, u, s) &= -u^+ + \lambda_1/v + \lambda_2(u - r(v) - g(s))/v \\ &= \begin{cases} (\lambda_2/v - 1)u + (\lambda_1 - \lambda_2 r(v) - \lambda_2 g(s))/v & \text{if } u \geq 0 \\ (\lambda_2/v)u + (\lambda_1 - \lambda_2 r(v) - \lambda_2 g(s))/v & \text{if } u < 0, \end{cases} \end{aligned} \tag{4.11}$$

with λ_1 and λ_2 the co-state variables associated to time and speed, respectively. The co-state variable λ_1 is a negative constant which depends on the available running time supplement [16].

It can be observed that the Hamiltonian is linear in the control u for both non-negative and negative control values, with coefficient $\lambda_2/v - 1$ for non-negative traction and λ_2/v for braking. Therefore, the optimal control structure can be split into five parts [16],

$$\hat{u}(s) = \begin{cases} u^{\max}(v(s)) & \text{if } \lambda_2(s) > v(s) & \text{(Maximum acceleration)} \\ r(v(s)) + g(s) & \text{if } \lambda_2(s) = v(s) & \text{(Cruising by partial traction)} \\ 0 & \text{if } 0 < \lambda_2(s) < v(s) & \text{(Coasting)} \\ r(v^{\max}(s)) + g(s) & \text{if } \lambda_2(s) = 0 & \text{(Cruising by partial braking)} \\ u^{\min}(v(s)) & \text{if } \lambda_2(s) < 0 & \text{(Maximal braking).} \end{cases} \tag{4.12}$$

The regime of cruising by partial traction corresponds to a unique optimal cruising speed v_c given as the implicit solution of [16]

$$v_c^2 r'(v_c) + \lambda_1 = 0. \tag{4.13}$$

The absolute value of $\lambda_1 < 0$ increases for shorter running time supplements, and therefore the optimal cruising speed is higher when less time supplement is available. In particular, the optimal cruising speed may be larger than the (local) speed limit depending on the specific conditions. Hence, the cruising speed by partial traction is given by $v(s) \equiv \min(v_c, v^{\max}(s))$, i.e., cruising at the speed limit is optimal for relative short time supplements. In contrast, cruising by partial braking only occurs at a speed limit $v^{\max}(s)$ during a downhill slope with gradient $g(s) \in \left[u^{\min} - r(v^{\max}), -r(v^{\max})\right]$ [16]. The cruising regimes may also be absent if the optimal cruising speed cannot be reached over relative short distances, depending on the stop distance and the speed limits.

The switching points in (4.12) depend on the co-state variable λ_2, which satisfies an adjoint ordinary differential equation defined by the negative partial derivative to speed of the Hamiltonian and the speed-dependent path constraints (4.4)–(4.5) [16],

$$\dot{\lambda}_2(s) = \frac{\lambda_1 + v\lambda_2 r'(v) + \lambda_2(u - r(v) - g(s))}{v^2} + \mu_1 u^+ + \mu_2, \tag{4.14}$$

where $\mu_1, \mu_2 \geq 0$ are Lagrange multipliers corresponding to the complementary slackness conditions of the path constraints $\mu_1(u^{\max}(v) - u) = 0$ and $\mu_2(v^{\max}(v) - v) = 0$, respectively. Hence, solving this optimal control problem requires solving a two-dimensional constrained boundary value problem of (v, λ_2) with boundary conditions $v(s_0) = v(s_1) = 0$ and none in λ_2, while depending on an implicit control function $u(v, \lambda_2)$ and an unknown cruising speed v_c (or λ_1). Algebraic formulae for the co-state λ_2 along track sections with constant gradient can be derived, which can be used to design efficient algorithms [1, 2, 18, 20]. In addition, constructive heuristic methods have been applied using the implicit knowledge of the optimal control structure (4.12). These solution methods are indirect methods in the sense that they are based on solving the derived optimality conditions from the Pontryagin's Maximum Principle. An alternative is given by direct solution methods that transcribe the continuous optimal control problem into a discrete nonlinear programming (NLP) problem by discretizing the state and control variables, the differential equations and the integral objective function. The resulting NLP problem can then be solved using efficient nonlinear optimization algorithms without a priori knowledge of the control structure [4]. In particular, pseudospectral methods have been developed for train trajectory optimization problems [16, 34, 37, 42]. The solutions so obtained can be checked to satisfy the optimality conditions, and in particular the optimal control structure (4.12) [16].

4.3.3 Illustrative Examples

This subsection illustrates the energy-efficient driving between two stops for several scenarios: long-distance versus short distance, the impact of a temporary speed restriction, and the effect of different running time supplements. The main results are summarized in Table 4.1.

Figure 4.3 from the previous section illustrates typical energy-efficient train trajectories for long and short distances. The static speed limit is 140 km/h in both cases. The applied running time supplements are 10% of the minimum running times. Figure 4.3 (left) shows the minimum-time (red) and the energy-optimal speed profile (blue) for a long distance train over a distance of 50 km. For the given running time supplement the optimal cruising speed is 132.2 km/h and the train starts coasting at 40.4 km, so that the train does not consume any traction energy in the last 9.6 km (19.2%). The coasting regime lasts for 325 s, which is 21.4% of the scheduled running time. Compared to the minimum time train control the energy-efficient train control thus applies a cruising speed below the speed limit and also has an additional coasting regime. The energy consumption reduces from 524.5 to 409.5 kWh, so that the energy-optimal driving strategy saves 21.9% energy with respect to the minimum running time.

Figure 4.3 (right) shows the minimum-time (red) and the energy-optimal speed profile (blue) for a local train over a short distance of 5 km. The optimal cruising speed is not reached within the distance and so the train accelerates to an optimal coasting point at 1.7 km reaching a maximal speed of 100.6 km/h after which the train starts coasting until the final braking regime. In this case, the train uses only maximal traction for the first part of the train run and then no longer requires traction energy for the remaining 3.3 km (66%). Note that also the minimum-time driving strategy cannot reach the speed limit within this distance and therefore accelerates to a speed of 129.3 km/h after which it already has to start braking to come to a standstill at the stop. The energy consumption reduces from 108.6 to 58.4 kWh, which is an energy saving of 46.3%.

Figure 4.4 (left) illustrates the energy-efficient speed profile for a long-distance train for various running time supplements, on a 50 km long track with a speed restriction of 125 km/h between 25 and 30 km. We here show both the impact of a speed restriction and varying running time supplements. Due to the speed restriction the minimum running time is 18 s longer than the long-distance case above without the speed restriction, since the train has to slow down its speed by 15 km/h during the speed restriction. The reduced speed and the reacceleration after the speed restriction leads to an increase in energy consumption of 12.7 kWh (2.4%) to 537.2 kWh for the minimum running time. The running time supplements are computed regarding this minimum running time.

In the case of 10% supplement, the optimal cruising speed is higher than the speed restriction and therefore the train has to reduce speed for this speed restriction. The optimal driving strategy now includes two coasting regimes, one before the speed restriction and one before the final braking at the end before the stop. At both sides

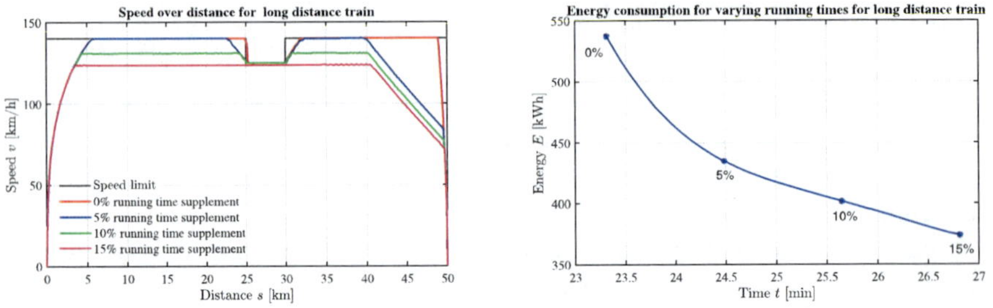

Fig. 4.4 The impact of a speed restriction (125 km/h) and varying running time supplements to a long-distance train: (left) speed-distance diagram for varying running time supplements and (right) the energy over running time diagram

of the speed restriction the train has a cruising regime with the same optimal cruising speed of 131.0 km/h, while during the speed restriction the train cruises at the lower restricted speed limit of 125 km/h. The switching points for both coasting regimes are determined implicitly by the unique cruising speed for the available running time supplement. The train re-accelerates with maximal tractive effort back to the optimal cruising speed after the train rear passed the speed restriction. With the 5% running time supplement the optimal cruising speed of 140.5 km/h exceeds the speed limit so that in this case the speed limits are applied in all cruising regimes. The difference with the minimum-time running is in this case only the coasting regimes before the speed restriction and before the stop. With 15% supplement the optimal cruising speed is 123.7 km/h, which is lower than the speed restriction so that in this case the restricted speed has no effect and the optimal driving strategy reduces to a single cruising and coasting regime like in the case of no speed restriction.

Figure 4.4 (right) illustrates the traction energy consumption as function of running time (supplement). When the amount of running time supplement increases the energy consumption goes down. In general, adding supplement in the beginning leads to more energy saving than increasing the supplement by the same amount later. Note that the last point corresponding to a train run with 15% running time supplement does not suffer from the energy loss due to the speed restriction, and therefore the energy saving with respect to the 10% point is more than from 5% to 10% in which cases the train has to reaccelerate after the speed restriction consuming extra energy.

It can be observed that much energy can be saved by adding even some small running time supplement percentage, as opposed to scheduling at the minimum running time. Likewise, a train will use much more energy when it runs as fast as possible in an attempt to recover a delay. Therefore, instead of running as fast as possible until a delay has been recovered, an energy-efficient delay recovery driving strategy is to recompute the optimal train trajectory for the remaining available running time supplement [35]. For instance, if the train scheduled with 10% supplement from Fig. 4.4 has a departure delay of 70 s then this reduces the remaining running time supplement from 10% to 5%. The optimal driving strategy then will change to the one corresponding to 5% running time supplement which will provide an on-time arrival time with the least energy consumption. Hence, in this case the optimal cruising speed will be increased to the speed limits and the coasting points start a bit earlier. The energy consumption will then increase by 33.1 kWh (8.2%) from 401.6 to 434.7 kWh.

4.4 Energy-Efficient Train Timetabling Over Multiple Stops

4.4.1 Problem Formulation

The previous section considered the problem of finding the energy-efficient train trajectory between two stops for a given scheduled running time. When designing a train schedule over a line with multiple stops, the total scheduled running time over the line may be given while the allocation of the running time supplements between the successive stops may still be optimized. This is in particular relevant for stops without sidings on a railway line between multi-platform stations where trains may meet and overtake. The total running time between the stations then consists of train runs over one or more stops plus the dwell times at the intermediate stops. A scheduling approach often used in practice is to consider each train run between two adjacent stops separately and allocating the same percentage of running time supplement to these segments.

In this section we consider the multi-stop energy-efficient train trajectory optimization problem. The aim is find the optimal distribution of running time supplements between successive stops of a single train in a corridor for a given total scheduled running time T, such that the total traction energy of the train trajectory over the entire line is minimized. This problem is an extension of the energy-efficient train control problem between two stops. It can also be viewed as an optimal scheduling problem that finds the optimal arrival and departure times at intermediate stops such that the total energy consumption of the train trajectories in the corridor is minimized for a given total scheduled running time T.

The (multi-stop) energy-efficient train trajectory scheduling problem aims to minimize the total traction energy of the train trajectory of a train over a corridor with n segments over given stop positions $(s_0, \ldots, s_n)$ with a total scheduled running time T. The optimal control problem is then given as

$$\text{Minimize} \int_{s_0}^{s_n} u^+(s)ds \tag{4.15}$$

under the constraints (4.2)–(4.5) and the boundary (and internal) conditions

$$t(s_0) = 0, \ t(s_n) = T, \ v(s_k) = 0, \quad k = 0, \ldots, n. \tag{4.16}$$

Note that the only difference with the EETC problem formulation from Sect. 4.3 is the extra stops (zero speed) at the $n-1$ intermediate positions $s_1, \ldots, s_{n_1}$, while the internal times $t(s_k)$ are free for these intermediate stops. The multi-stop train trajectory optimization will therefore also determine the optimal intermediate arrival times $t(s_k)$. Note that the dwell times at the stops have been discarded, so the total running time T has been reduced by the sum of intermediate dwell times, which can be added to the stops after the optimization of the running times. The dwell times can also be included explicitly by defining free arrival and departure times at each stop that are connected by a fixed given dwell time in between [35, 36].

4.4.2 Optimality Conditions

The necessary optimality conditions are exactly the same as the solution of the two-stop case given in Sect. 4.3.2, i.e., the optimal control structure (4.12) still holds for the multi-stop case, including the implicit expression of the optimal cruising speed (4.13) and the dynamic equation (4.14) for the co-state λ_2 [30]. In particular, it can be proved that the optimal cruising speed is unique over the entire line when the arrival and departure times of intermediate stops are not constrained by restricted time windows [17]. Note that this problem is a special case of the train trajectory optimization problem between two stops with fixed scheduled running time, where the speed bound constraint (4.4) includes zero upper bounds at discrete intermediate points. As was illustrated in Fig. 4.4, the optimal cruising speed was uniquely determined by (4.13) resulting in an equal cruising speed at both sides of a speed restriction. The knowledge that the optimal cruising speed is unique over a corridor can be used for designing efficient algorithms [17, 30].

Of course, the internal boundary conditions force the train trajectory to zero speed at the intermediate stop positions, by which the speed profile looks like several train trajectories over successive stops. However, these train trajectories are dependent in the sense that the overall running time supplement is optimally distributed between all stops. This causes the same optimal cruising speed on each segment of the corridor and also determines the exact switching points between the driving regimes on each of them.

If an intermediate stop has a fixed target arrival or departure time, then the problem is split into two separate train trajectory optimization problems with the fixed event time at this stop as the final condition for the first problem and as initial condition for the second problem. This will reduce the flexibility of the optimization and thus result in a larger total energy consumption over the corridor. In particular, when the times at all intermediate stops are scheduled in advance before the train trajectory optimization then the energy consumption can only be optimized for the fixed scheduled running times between successive stops. The optimal scheduling method from this section, on the other hand, would exchange running time supplements between segments such that any supplement is added where needed the most.

4.4.3 Illustrative Example

This section illustrates the energy-efficient trajectory optimization and scheduling strategies over multiple stops with a case study for an intercity (IC) train from Utrecht Central (Ut) to Arnhem Central (Ah) in the Netherlands, with three intermediate stops in Driebergen-Zeist (Db), Veendendaal-De Klomp (Klp) and Ede-Wageningen (Ed). The speed limit is 140 km/h and the total distance is 60 km, with the intermediate stops at 10, 33 and 40 km. So this line has two shorter station distances of 10 and 7 km, and two longer station distances of 23 and 20 km. The minimum running time is computed as 1930.4 s, consisting of the successive minimum running times 354.0 s (Ut-Db), 688.3 s (Db-Klp), 276.9 s (Klp-Ed) and 611.2 s (Ed-Ah). The total running time supplement over Ut–Ah is assumed to be 15%, i.e., 290 s, resulting in a scheduled running time of 2120 s (excluding the dwell times at the stops). Note that nowadays timetables in the Netherlands are calculated with a precision of 1/10 min.

The minimum-time and energy-efficient scheduling strategies are compared to other scheduling strategies that are often used in practice. In total, this section considers the following five scheduling strategies over multiple stops:

1. Minimum-time train trajectory,
2. Energy-efficient train trajectory with optimal distribution of supplements,
3. Uniform schedule with equal percentage of running time supplement (15%),
4. Tightened schedule with most supplement added to the end,
5. Actual schedule with oscillating supplement percentages.

Table 4.1 summarizes the main results of these strategies, while Table 4.2 shows the optimal running time distribution and the maximum speed for each segment.

First, we focus on the energy-efficient train trajectories. Figure 4.5 (left-top) shows the energy-efficient speed profiles (in green). The 2nd and 4th long segments show identical optimal cruising speeds of 131.2 km/h, while the coasting starts later for the longer 2nd segment. The 1st and 3rd short segments are too short to reach the optimal cruising speed and therefore the train accelerates to an optimal coasting point with associated coasting speed after which it starts coasting. The coasting speed depends on the stop distance and is higher for the longer 1st segment where switching occurs at speed 118.8 km/h, while on the 3rd segment switching to coasting occurs at a speed of 104.0 km/h.

Figure 4.5 (right-bottom) shows the optimal relative running time supplements in percentage of the associated running times, which are deceasing for longer stop distance. Hence, when distributing a fixed amount of running time supplement over a schedule with multiple stops, allocating relative more running time supplement to shorter distances will lead to more energy saving than a uniform distribution. For

Table 4.1 Main results of the different scenarios

Scenario	Figures	Running time [s]	Supplement [%]	Energy [kWh]	Energy saving [%]
MTTC long distance	4.3	1380.7	0	524.5	–
EETC long distance	4.3	1518.8	10.0	409.5	21.9
MTTC short distance	4.3	222.1	0	108.6	–
EETC short distance	4.3	244.3	10.0	58.4	46.3
MTTC speed restriction	4.4	1398.7	0	537.2	–
EETC 5% supplement	4.4	1468.6	5.0	434.7	19.1
EETC 10% supplement	4.4	1538.5	10.0	401.6	25.2
EETC 15% supplement	4.4	1608.5	15.0	373.8	30.4
MTTC Ut–Ah	4.5	1930.4	0	602.1	–
EETC Ut–Ah tightening	4.5	2120.0	15.0	400.0	33.6
EETC Ut–Ah uniform	4.5	2120.0	15.0	365.6	39.3
ETTC Ut–Ah practice	4.5	2120.0	15.0	391.4	35.0
ETTC Ut–Ah optimal	4.5	2120.0	15.0	364.7	39.4

Legend MTTC = mimimum time train control, EETC = energy-efficient train control

Table 4.2 Running time supplements and maximum speed on the successive segments over the line Ut-Ah for five scheduling strategies

Segment	Ut–Db	Db–Klp	Klp–Ed	Ed–Ah
Supplement [s, %]				
MTTC	0 (0%)	0 (0%)	0 (0%)	0 (0%)
ETTC tightening	17.7 (5.0%)	34.4 (5.0%)	13.8 (5.0%)	223.6 (36.6%)
EETC uniform	53.1 (15.0%)	103.2 (15.0%)	41.5 (15.0%)	91.7 (15.0%)
EETC practice	152.8 (43.2%)	55.9 (8.1%)	69.6 (25.1%)	11.2 (1.8%)
EETC optimal	64.9 (18.3%)	89.2 (13.0%)	50.9 (18.4%)	84.6 (13.8%)
Maximal speed [km/h]				
MTTC	140.0	140.0	139.2	140.0
ETTC tightening	134.2	140.0	121.4	102.8
EETC uniform	122.0	128.1	107.7	129.0
EETC practice	96.5	140.0	98.5	140.0
EETC optimal	118.8	131.2	104.0	131.2

instance, the running time supplement over the shortest distance of 7 km (Klp–Ed) is 18.4% (51 s), while this is 13.0% (89.2 s) for the longest distance of 23 km (Db–Klp). Note that in terms of absolute running time supplements, those for the longer distance still may be longer, see Table 4.2. Overall, the energy-optimal train trajectory and schedule over the corridor with 15% total running time supplement saves 237.4 kWh (39.4%) energy consumption with respect to the minimum running time.

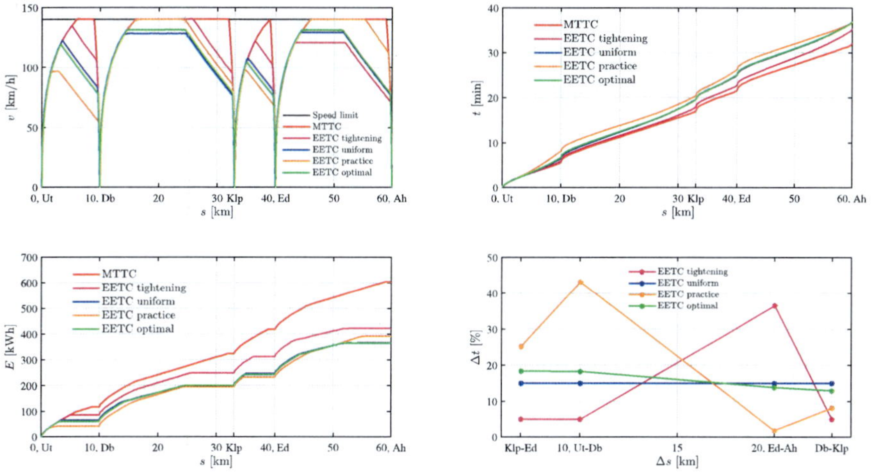

Fig. 4.5 Train trajectories for a train over multiple stops (Ut, Db, Klp, Ed, and Ah) for various driving strategies (minimum time, energy-optimal, uniform, tightening, practice): (left-top) speed-distance, (right-top) time-distance, (left-bottom) energy-distance, and (right-bottom) relative running time supplement over distance

Figure 4.5 also compares the results for the various scheduling strategies used in practice. The uniform schedule illustrates the deviation of the energy-optimal train trajectory from a uniform supplement percentage of 15%, i.e., providing a smaller supplement at the short stop distances and a larger one at the long stop distances. In the uniform case, the cruising speeds over the longer distances are lower than the optimal cruising speed corresponding to the longer running times on these segments, and they also vary with a lower cruising speed for the longer distance, 128.5 km/h on Db–Klp (23 km) and 130 km/h on Ed-Ah (20 km). On the short distances the maximal speeds obtained before coasting are higher than the optimal ones due to the shorter running times, 122 km/h on the 1st segment and 108 km/h on the 3rd. Still the increase in total energy consumption is negligible with less than 1 kWh (0.24%).

The tightened schedule was used in the period 2008–2015 in the Netherlands with the aim to obtain more robust schedules in the sense that the majority of the running time supplement is allocated before stations where punctuality is measured, which in this case is on the last segment of the corridor before Arnhem Central. Then, a delay on any of the segments on the corridor can be recovered as much as possible as opposed to 'wasted' supplements at the early segments when a delay occurs at the later segments. In this case, the first three segments still obtained (the minimum required) 5.0% running time supplement, while 36.6% is allocated at the last segment, see Table 4.2. As a result the maximal speeds obtained before coasting at the 1st and 3rd short distance segments are much higher then the optimal coasting speeds, and in the 2nd segment the train cruises at the speed limit of 140 km/h. In contrast, the cruising speed in the last segment is much lower with only 102.8 km/h. As a result, much energy is lost over specifically the two short distances and this can hardly be compensated by the excessive supplement on the last long distance segment. This strategy uses 35.3 kWh (9.7%) more energy than the optimal strategy.

The current actual schedule (no longer designed according to the tightened strategy) uses significantly higher percentages at the short distances of 43.2% and 25.1% at Ut–Db and Klp–Ed, respectively, and lower percentages at the long distances of 8.1% on Db–Klp and only 1.8% on Ed–Ah. The planners added extra running time supplement before Db and Ed, at the cost of the supplement before Klp and in particular Ah. As a result the reached speeds at the short distances are lower with longer coasting regimes, while on Ut-Db even a short cruising regime at 96.5 km/h occurs to avoid that the speed becomes too slow at the end of the coasting regime. On the long distances cruising at the speed limit is applied with in particular only a short coasting regime on the last long distance segment. In this case, the additional energy consumption on the corridor is 26.7 kWh (7.3%).

4.5 Energy-Efficient Timetabling of Multiple Trains Over a Corridor

4.5.1 Problem Formulation

Unnecessary loss of energy occurs if a train must brake due to a conflicting preceding train and then needs to re-accelerate when the route ahead is available again. Not only does the re-acceleration cause additional energy consumption compared to a conflict-free train run, but the train also has to run faster to compensate for the delay caused by the unplanned braking and possible waiting before a signal until the train is allowed to proceed again. Train path conflicts may be caused by inaccurate timetabling or by schedule deviations during operations either by a delayed preceding train or by early running, when two trains want to occupy a shared track at the same time. In practice the railway signalling systems will intervene by which one of the two trains will have to slow down. This train will then be delayed and allocate tracks for a longer time than scheduled, which may cause a further cascade of path conflicts and delays to later trains in case of heavily occupied networks.

The railway timetable therefore must provide conflict-free train paths such that each train should be able to run according to its scheduled train trajectory without route occupation conflicts. This is also called a green wave in the sense that the trains will not meet restricted signals due to conflicting trains if they all adhere to their schedule [8, 34]. In addition, a railway timetable must be robust in the sense that it should contain some buffer time between successive train paths such that a slight deviation from the schedule will not immediately lead to a route occupation conflict [15]. In the case of bigger disturbances traffic management must detect possible conflicts and resolve them proactively to maintain a conflict-free traffic plan to avoid a waste of energy and track capacity [25].

In the previous sections the focus was on scheduling a single train trajectory between stops or over corridors. From the infrastructure perspective, the train trajectory allocates the track sections on a route exclusively to a train over certain time slots during which they are blocked to other trains. These blocking times depend on the speed profile of a train and the signalling system applied that guarantees safe train separation. In general, the railway infrastructure is partitioned into block sections that may contain only one train at a time. The signalling system dictates at which location a brake indication has to be given when approaching an occupied block. For conflict-free train running the block should be clear before reaching the brake indication point to that block. Otherwise, the train has to brake and thus deviate from its planned train trajectory.

Blocking times are computed using blocking time theory [22, 23]. The blocking time for a given block consists of the sum of six time components: the setup time, sight and reaction time, approach time, running time, clearing time, and release time. The setup time is the required time to set and lock the route in the block. The sight and reaction time represents the required time to respond to a brake indication in case of a restricted movement authority. The approach time models the running time over the approach distance from the brake indication point to the actual block. The running time is the time within the block. The clearing time is the running time over the train length at the end of the block until the complete train has left the block. And finally, the release time is the time to release the route and signals to be used by the next train. The approach, running and clearing time depend on the train speed and therefore the train trajectory, while the other three time components are often defined as time parameters conditional on the given infrastructure conditions.

Figure 4.6 illustrates the blocking time components for a single block section of a running train under three-aspect two-block signalling, where lineside signals indicate the movement authority using three aspects: stop (red) indicating that the train should stop before the block, approach (yellow) dictating to reduce speed and prepare to stop before the next signal, and clear (green) meaning that the train can proceed with the track speed [14]. In this case the brake indication is thus given by the approach signal. If the route in a block includes switches then all sections in the block are locked simultaneously but they may be released in parts to allow the switches to be set for another route, which is called the sectional-release route-locking principle, see the blocks between the 2nd and 3rd signal in Fig. 4.7. Note that the blocking time exceeds the physical occupation time of the block due the signalling constraints that require a train to start braking when it approaches a closed block and additional system times to set up and release the route.

Blocking times enrich the traditional time-distance diagrams by including the time slots that the successive blocks are blocked for a specific train path to allow conflict-free operation. With this additional information the infrastructure occupation of a train path operating under fixed-block signalling takes the form of a blocking time stairway, see Fig. 4.7. A conflict-free timetable should contain 'white space' between the blocking times of all successive trains. As a result, conflicts are now easily visualised by overlapping blocking times, which indicate that a train needs to reserve a block while it has not yet been released by the previous train, see Fig. 4.7

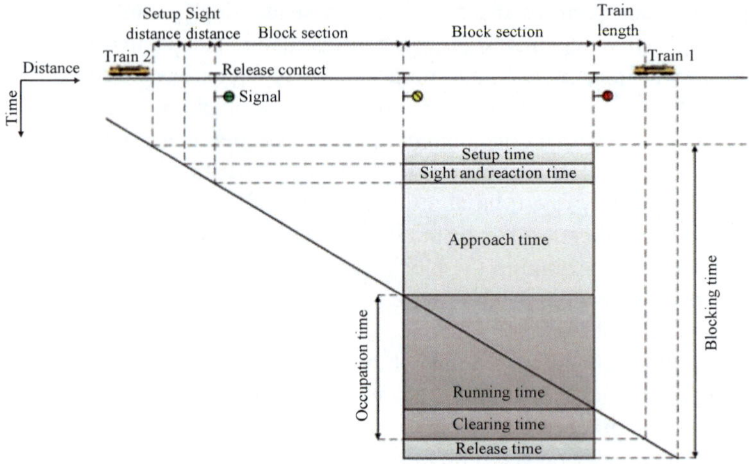

Fig. 4.6 Blocking time components

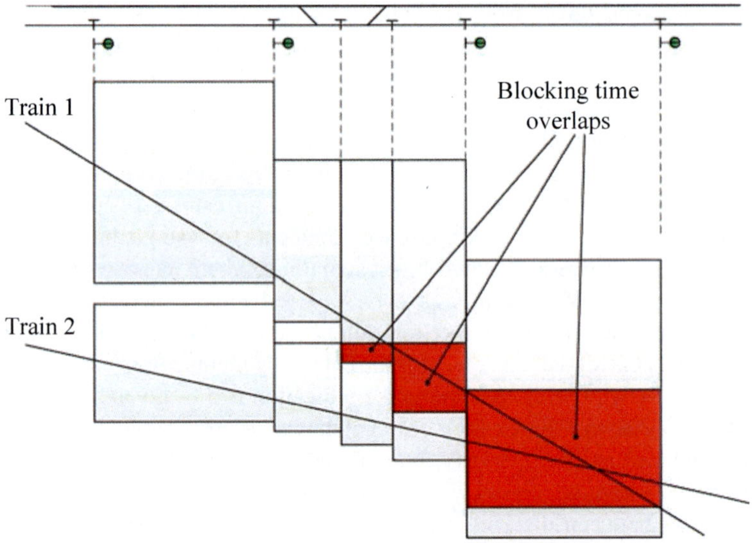

Fig. 4.7 Conflict detection by overlapping blocking time stairways

where two successive train paths are conflicting in the last two blocks. In addition, by compressing the blocking time stairways over a corridor the minimum line headway times can be calculated corresponding to critical blocks between the successive train paths where adjacent blocking times are touching each other. If the scheduled train paths respect these minimum headway times at the line level then they will be conflict-free, including any running time differences between successive trains for heterogeneous traffic. A conflict-free timetable should include some buffer time on top of the minimum line headway times to be robust to small train path deviations. The compressed blocking time stairways of all train paths scheduled in a given time period without any buffer time at the critical blocks give the infrastructure occupation of the associated timetable structure, which is a measure of the used capacity [15].

In the following, we assume that the desired scheduled departure and arrival time at respectively the first and last station of a corridor are given, and hence also the scheduled running time for each train on the corridor. The problem is then an extension to the one from the previous section to find the energy-efficient train trajectories of all trains over the corridor such that they do not have conflicting infrastructure occupation. This problem can be modelled as a multi-train trajectory optimization problem [35, 36], where the event times of m successive trains at certain positions are restricted by minimum headway times, as follows.

Let n_i be the number of successive train runs of train $i \in \{1, \ldots, m\}$ and $s_{ik} \in S_i = \{s_{i0}, \ldots, s_{in_i}\}$ the stop positions of train i. Note that the trains may have different stop positions. Denote common locations of two successive trains i and j (in this order) that require a headway constraint by $(s_{ik}, s_{jl}) \in S_{ij} \subseteq S_i \times S_j$ for $i \neq j$ with corresponding headway time h_{ikjl}. The timing points for headway constraints can also be generalized to fixed locations other than the stop positions, such as signal positions or station locations for non-stopping trains. The state and control variables as well as the resistance, speed and gradient profiles and other parameters for a train i are indicated with an index i. Finally, for each train i the scheduled departure time d_i at the initial stop position s_{i0} and the scheduled arrival time a_i at the final stop position s_{in_i} is given. Then the multi-train trajectory optimization problem can be defined as

$$\text{Minimize} \sum_{i=1}^{m} \int_{s_{i0}}^{s_{in_i}} u_i^+(s)ds \tag{4.17}$$

subject to the following constraints for all trains $i \in \{1, \ldots, m\}$,

$$\dot{t}_i(s) = 1/v_i(s) \tag{4.18}$$
$$\dot{v}_i(s) = (u_i(s) - r_i(v) - g_i(s))/v_i(s) \tag{4.19}$$
$$0 \leq v_i(s) \leq v_i^{\max}(s) \tag{4.20}$$
$$u_i^{\min}(v_i) \leq u_i(s) \leq u_i^{\max}(v_i), \tag{4.21}$$

the boundary and internal conditions

$$t_i(s_{i0}) = d_i, \ t_i(s_{in_i}) = a_i, \ v_i(s_{ik}) = 0, \quad s_{ik} \in S_i, \tag{4.22}$$

and the pairwise headway time constraints linking the train trajectories at fixed locations

$$t_j(s_{jl}) - t_i(s_{ik}) \geq h_{ikjl}, \quad (s_{ik}, s_{jl}) \in S_{ij}, \tag{4.23}$$

for all trains $i, j \in \{1, \ldots, m\}$. Recall that we considered fixed event times at the beginning and end of the corridor and so also the sequence orders are fixed between the trains over the corridor including possible order changes at overtaking locations. In a more general timetable optimization problem the order may still have to be decided which will complicate the headway constraints (4.23).

Note that trains can take different routes, so that formally the independent variable s can be different for each train i, and therefore also the gradient $g_i(s)$ and speed limits $v_i^{\max}(s)$ depend on train i. For the modelling of different routes using multiple-phase optimal control, see Wang and Goverde [35, 36].

4.5.2 Solution Procedure

The necessary optimality conditions from Sect. 4.4 are still valid for each train trajectory in the multi-train multi-stop train trajectory optimization problem, and in particular the optimal control structure (4.12) still holds with a unique cruising speed per train over the entire corridor. The headway time constraints between the time-distance paths of pairs of trains at fixed locations provide additional timing constraints that depend on both train trajectories, as opposed to simple fixed timing or speed constraints for single train trajectories. If the inequalities (4.23) are not active then all trains will be able to run according to the energy-efficient single-train trajectories. Only in case of conflicts the headway time constraints become active and the train trajectories have to be jointly optimized to satisfy these constraints.

The general procedure to find energy-efficient train trajectories of multiple trains over a corridor can be given as follows [35, 36]:

1. Solve the energy-efficient train trajectory optimization problem (over two or more stops) for each train in the corridor for fixed departure and arrival time at the corridor ends.
2. Compute the corresponding blocking time stairways for all trains and check for overlapping blocking times (i.e., conflicts).
3. Resolve the conflicts by a multi-train trajectory optimization problem over the conflicting trains considering headway constraints.

We can distinguish three different situations for scheduling trains over a corridor:

- Double-track lines with one running direction per track,
- Double-track lines with one running direction per track and intermediate overtaking stations or sidings,
- (Partially) single-track lines with opposite trains meeting at passing stations or sidings.

Also combinations are possible with in particular multiple-track lines, and two railway lines may merge at a junction so that trains from different lines are combined on the track after the junction, or in the opposite direction, successive trains can divert into different directions after a diverging junction. The basic principles still apply to these situations.

If the interval between successive scheduled departure times at the origin station of a corridor exceeds the minimum line headway times then the associated blocking time stairways will not overlap and the train trajectories will be conflict-free. An exception occurs if the railway line allows overtakings (the 2nd situation) or meetings between opposite trains (the 3rd situation), in which cases the train trajectories must be synchronized at the meeting or overtaking points.

Since railway timetabling is an NP-hard problem, Goverde et al. [15] proposed a three-level framework where first a conflict-free timetable on the network level is computed using a micro-macro iterative approach, after which energy-efficient train trajectories are embedded over the corridors. The microscopic level computes the running and blocking times and checks for conflicts and acceptable infrastructure occupation using blocking time theory, while the macroscopic level optimizes the event times at the main stations considering optimal and robust travel times [5]. Then at the third fine-tuning level the corridors between the main stations are optimized for fixed target times at the corridor ends using train trajectory optimization. First, the non-stop intercity train trajectories are computed using the train trajectory optimization problem over the entire corridor, like the model from Sect. 4.3. For the local train trajectories over the corridors, the impact of stochastic intermediate dwell times were considered using a stochastic model formulated as a multi-stage multi-criteria decision problem that was solved by dynamic programming. The model from the current section is an alternative approach to solve the energy-efficient timetable problem over the corridors. Moreover, this model can also be embedded at the microscopic level. The third level can then be discarded unless the stochastic dwell times should be considered. A simpler rapid running time calculation model was developed for the microscopic level that determined the timetable (cruising) speed to cover the scheduled running time without considering coasting. Since in this iterative framework train trajectories have to be (re)computed for many lines over a large-scale network rapid computation time is crucial.

4.5.3 Illustrative Examples

In this section we provide two examples corresponding to the situations of a double-track line [28] and a double-track line with an overtaking [36]. For the case of single-track lines, see Wang and Goverde [35, 36].

The first case study considers the Dutch corridor of about 18.5 km between the main stations Arnhem Central (Ah) and Nijmegen (Nm), with the intermediate stations Arnhem Zuid (Ahz), Elst (Est) and Nijmegen Lent (Nml). A long-distance intercity (IC) train only stops at the main stations Ah and Nm, while the short-distance regional (RE) train stops at all stations with 42 s dwell time. Both train types share the same route over the corridor, except at the first and last block section due to different track and platform use at Ah and Nm, so we do not consider overtaking on the railway line between these stations. For details of the rolling stock characteristics, blocking time parameters, and the track characteristics such as gradients, speed limits and signal positions, see Scheepmaker and Goverde [28].

The aim of this case study is to find an energy-efficient timetable for alternating intercity and regional trains from Ah to Nm that repeat at a regular interval of 15 min, so with a frequency of 2×4 trains per hour. A minimum running time supplement of 8% is required for both trains, and a minimum buffer time of 30 s must be included over the minimum line headway time between the trains. The quality of the timetable can be measured by four indicators: total scheduled running time (including intermediate dwell times) and energy consumption of both train types, the infrastructure occupation, and the total buffer time. Larger running times imply lower energy consumption, because more supplement is available for energy-efficient driving. On the other hand, the infrastructure occupation depends on the running time difference between the IC and RE trains, and therefore on the relative allocation of running time supplement to the two trains over this corridor. Finally, higher infrastructure occupation means less buffer time and therefore a lower stability regarding delays. The infrastructure occupation can be minimized by homogenizing the two trains, i.e., slow down the IC trains by adding a large running time supplement, or speeding up the RE trains by including only the minimum supplement. However, this will have a negative affect on the running time for the IC train or on the energy consumption of the RE train. An optimal solution can be found by considering a multi-objective optimization problem, and in particular a weighted sum of the four indicators [28]. The optimal solution allocates 12% running time supplement to the IC train and 9.5% to the RE train. The total buffer time is 4 min and the infrastructure occupation is 73.3%. Figure 4.8 illustrates the compressed blocking time diagram for the energy-efficient driving strategies. The shown blocking times are extended with the 30 s required minimum buffer time (dark areas). The minimum line headway time is $156 + 30$ s minimum buffer time from the IC to the RE train, and $504 + 30$ s from the RE to the IC train. A regular interval timetable can be obtained by dividing the extra 180 s buffer time between the train pairs, such that the second IC train departs at 15 min (in the next cycle). Possible departure times from Ah could then be the IC at 0 s and the RE at 240 s repeating every 15 min (900 s), with resulting buffer times of 94 s and 156 s, respectively.

The second case study considers the 50 km long Dutch corridor between the main stations 's-Hertogensbosch (Ht) and Utrecht Central (Ut), with six intermediate stations Zaltbommel (Zbm), Geldermalsen (Gdm), Culemborg (Cl), Houten Castellum (Htnc), Houten (Htn), and Utrecht Lunetten (Utl). Gdm is an overtaking station. The considered periodic timetable includes four trains per 30 min: two IC trains stopping only in Ht and Ut, one RE train from Ht to Ut that is overtaken by an IC train at Gdm, and another RE train that merges into the corridor at Gdm and runs to Ut after the other IC train. The scheduled departure and arrival times at the start and end stations are given and so are the original arrival and departure times at all intermediate stops for the RE trains, see Table 4.3. Note that the trains of equal type do not follow a strict departure interval of 15 min, but a 14–16 min interval. In the longer interval additional freight paths can be operated, which are not considered here. The dwell times on the short stops of the RE trains should be within 30 and 60 s, and at the overtaking station Gdm between 180 and 360 s. For an overtaking a default headway time of 120 s applies both between arrival of the stopping train and the passing IC

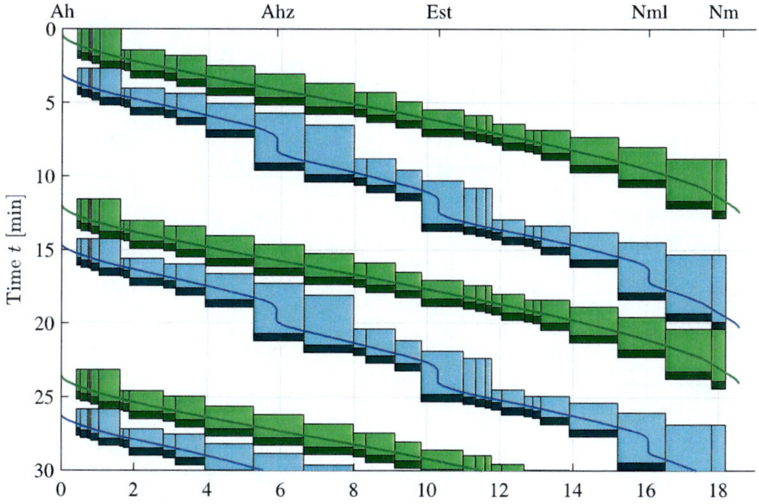

Fig. 4.8 Compressed blocking time diagram with energy-efficient train trajectories for optimized running time supplements of an IC (green) and RE train (blue)

train, and from the passing IC train to the departure of the stopping train. For details about the rolling stock and track characteristics, see Wang and Goverde [36].

The aim of this case study is to find an energy-efficient timetable over the corridor for the four trains per hour, while the start and end times are fixed. The results are summarized in Table 4.3. Figure 4.9 (left) illustrates the results of the train trajectory optimization for all four trains separately, with the corresponding blocking times. The two IC trains show the energy-efficient train trajectory between two stops, which are the same as both have the same scheduled running time of 28 min. The train trajectory optimization of the RE trains includes optimization of the arrival and departure times at the intermediate stops, see Table 4.3. In particular, the 2nd RE train running from Ht to Ut arrives 1 min later at Gdm compared to the original timetable and departs 1 min earlier, thus reducing the dwell time at Gdm from 5 to 3 min. Therefore, also the remaining running time from Gdm to Ut increases from 15 to 16 min, by which the train trajectory of this RE train is different from the one that starts from Gdm at the original fixed departure time. The energy saving is 21.8% for the RE train Ht-Ut and 17.3% for the other RE train Gdm-Ut compared to the energy-efficient train trajectories that can be obtained while sticking to the original timetable at the intermediate stops. The energy consumption of the IC trains in the original timetable is also computed for the energy-efficient train trajectories, although the original speed profiles were probably different.

Figure 4.9 (left) shows overlapping blocking time stairways after Gdm between the RE train and the IC train that should overtake this RE train at Gdm (the red blocking times). Since the dwell time of the RE train is now 3 min, it does not satisfy the default headway times of 2 min before and after passage by the IC. Therefore, we apply a multi-train trajectory optimization for these two trains with additional headway constraints of 120 s at Gdm between the arrival time of the RE train and

Table 4.3 The original and optimized timetables for the double-track corridor Ht-Ut

Train	Event	Ht	Zbm	Gdm	Cl	Htnc	Htn	Utl	Ut	Energy [MJ]	Saving
Original timetable											
RE1	A	–	–	–	12:30	18:30	22:30	26:30	32:00	0.3966	–
	D	–	–	07:00	13:00	19:00	23:00	27:00	–		
RE2	A	–	11:30	18:00	28:30	34:30	38:30	42:30	48:00	0.7342	–
	D	02:00	12:00	23:00	29:00	35:00	39:00	43:00	–		
IC1	A	–	–	–	–	–	–	–	37:00	1.2394	–
	D	09:00	–	–	–	–	–	–	–		
IC2	A	–	–	–	–	–	–	–	51:00	1.2394	–
	D	23:00	–	–	–	–	–	–	–		
Single-train trajectory optimization											
RE1	A	–	–	–	13:00	20:00	22:48	27:18	32:00	0.3281	17.3%
	D	–	–	07:00	13:30	20:30	23:18	27:48	–		
RE2	A	–	11:42	19:00	28:30	36:00	38:48	43:12	48:00	0.5741	21.8%
	D	02:00	12:12	22:00	29:00	36:30	39:18	43:42	–		
IC1	A	–	–	–	–	–	–	–	37:00	1.2394	0%
	D	09:00	–	–	–	–	–	–	–		
IC2	A	–	–	–	–	–	–	–	51:00	1.2394	0%
	D	23:00	–	–	–	–	–	–	–		
Multi-train trajectory optimization											
RE2	A	–	12:00	20:00	30:00	36:42	39:30	43:00	48:00	0.6355	13.4%
	D	02:00	12:30	24:00	30:30	37:12	40:00	43:30	–		
IC1	A	–	–	–	–	–	–	–	37:00	1.2394	0%
	D	09:00	–	–	–	–	–	–	–		

A arrival time [mm:ss]; *D* departure time [mm:ss]

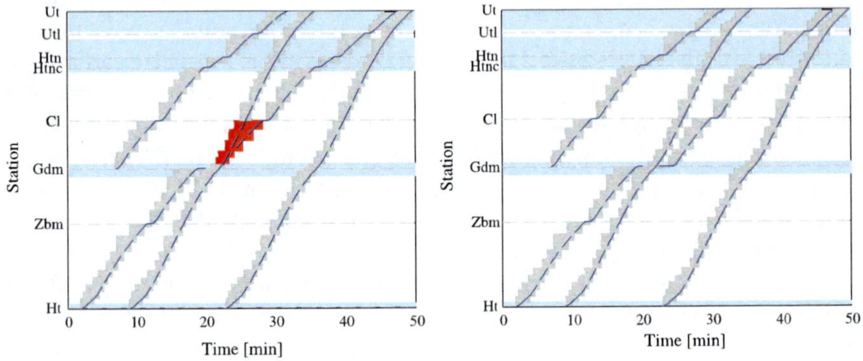

Fig. 4.9 Blocking time diagrams of energy-efficient train trajectories with conflict indicated by the red overlapping blocking times (left) and resolved conflict (right). *Source* [36]

the passage of the IC train, and likewise between the latter and the departure time of the RE train. The result is illustrated in Fig. 4.9 (right) that shows a conflict-free timetable. The arrival time of the RE train in Gdm is now another 1 min later and the departure time 2 min later, with the minimal dwell time of 4 min that is required by the two headway times. This will provide the maximal running time supplement for the RE over the corridor. The energy-efficient train trajectory of the IC train did not change, so that it was optimal to adjust the train trajectories from the RE train before and after the passage time of the IC train, which happens to be at 22 min in the basic half hour period. The energy saving of the RE train reduced to 13.4% to allow a conflict-free timetable.

Figure 4.10 illustrates the speed profiles of the various energy-efficient train trajectories. The speed profile for the IC train obtained from the single-train trajectory optimization (4th plot) is the same as the one obtained from the multi-train trajectory

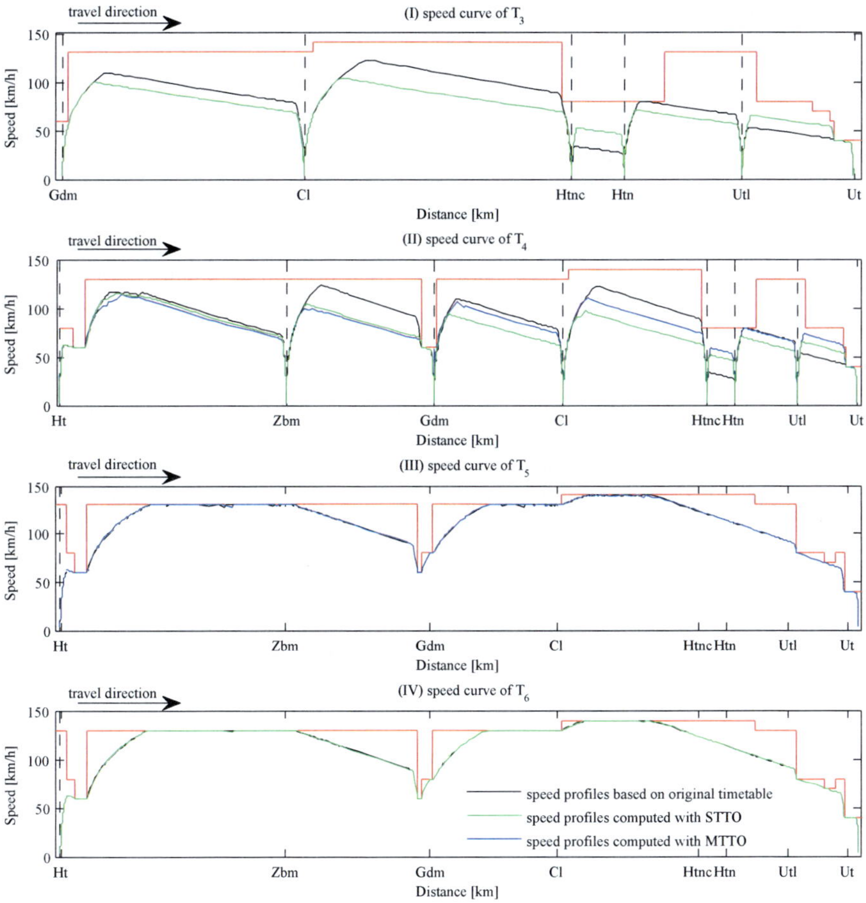

Fig. 4.10 Speed profiles for the trains on the double-track corridor Ht-Ut. *Source* [36]

optimization (3rd plot). For the RE train Ht-Ut (2nd plot) three different speed profiles can be observed. For the original timetable the speeds are higher before Gdm as the scheduled running time from Ht to Gdm is shorter than in the optimal timetables. The original scheduled running time over Gdm-Ut is 1 min larger than in the solution from the multi-train trajectory optimization. However, the latter optimizes the arrival and departure times at the intermediate stops, which results in a different allocation of the supplements between the various stops. The speeds of the optimized solution are still lower on the two longer runs Gdm-Cl and Cl-Htnc corresponding to 30 and 42 s more running time, while on the short stretches after Htnc they are higher than the very slow speeds resulting from the original timetable, corresponding to reduced running times by 72 s, 30 s, and 30 s. In particular, the running time over the short stretch from Htnc to Htn was reduced by 34%.

4.6 Conclusions

This chapter showed how energy-efficient train trajectory optimization can be incorporated in railway timetabling. In particular, the scheduled arrival and departure times at stations as well as passage times of non-stop trains at stations or other timing points should be based on realistic speed profiles that include the impact of the allocated running time supplement. Moreover, for saturated railway networks it is essential that the assumed driving behaviour in the running time calculations from the planning phase are consistent with the actual driving strategies in actual train operation. Otherwise, train path conflicts may arise during operation while the plan was 'proven' to be feasible.

Energy-efficient train operation must be supported by a railway timetable that allows for energy-efficient driving. This means that the running time supplements must be allocated between the successive stops in such a way that train operation can be energy-efficient, and no energy is lost by drivers who try to adhere to scheduled arrival times that are scheduled in a naive way. The latter may occur for instance by short stretches with relatively much running time (supplement) due to rounding to full minutes, or scheduling much supplement just before main stations to improve punctuality statistics. Scheduling in a higher precision than full minutes is recommended to provide improved information to drivers or Automatic Train Operation. For instance, since a few years the Netherlands Railways plan their timetable with a precision of a tenth of a minute.

The successive speed profiles should be predictable and drivable. Energy-efficient speed profiles have equal cruising speeds over successive train runs, as opposed to varying timetable speeds between timing points due to scheduling of event times that are not based on valid speed profiles, such as the typical timetabling practice based on minimum running times plus some percentage or fixed amount of running time supplement, and this rounded to full minutes. It is useless to provide training for drivers to run in an energy efficient way or to develop advanced algorithms for

Driver Advisory Systems or Automatic Train Operation, if the timetable prevents energy-efficient driving [26].

Moreover, timetables must be conflict-free and robust so that minor deviations from a train path do not immediately lead to conflicts and delay following trains, as this will also result in unnecessary braking and re-acceleration and therefore in loss of energy. Energy-efficient train timetabling thus also calls for a microscopic approach where the constraints from the signalling system are realistically modelled in blocking times and resulting infrastructure occupation, which enables conflict detection and resolution between the scheduled train trajectories to guarantee conflict-free timetables.

The mathematical models and algorithms are available to develop performance-based energy-efficient train timetabling, including microscopic train trajectory optimization and conflict detection and resolution [15]. In particular, energy-efficient train trajectory optimization should become the standard running time calculation method in any timetabling design tools.

References

1. Albrecht AR, Howlett PG, Pudney PJ, Vu X, Zhou P (2016) The key principles of optimal train control-part 1: formulation of the model, strategies of optimal type, evolutionary lines, location of optimal switching points. Transp Res Part B Methodol 94:482–508
2. Albrecht AR, Howlett PG, Pudney PJ, Vu X, Zhou P (2016) The key principles of optimal train control-part 2: existence of an optimal strategy, the local energy minimization principle, uniqueness, computational techniques. Transp Res Part B Methodol 94:509–538
3. Albrecht T, Oettich S (2002) A new integrated approach to dynamic schedule synchronization and energy-saving train control. In: Allen J, Hill RJ, Brebbia CA, Sciutto G, Sone S (eds) Computers in railways VIII. WIT Press, Southampton, UK, pp 847–856
4. Betts JT (2010) Practical methods for optimal control and estimation using nonlinear programming. Advances in design and control. SIAM, Philadelphia, PA, USA
5. Bešinović N, Goverde RMP, Quaglietta E, Roberti R (2016) An integrated micro-macro approach to robust railway timetabling. Transp Res Part B Methodol 87:14–32
6. Binder A, Albrecht T (2013) Timetable evaluation and optimization under consideration of the stochastical influence of the dwell times. In: Proceedings of the 3rd international conference on models and technologies for intelligent transportation systems, pp 471–481
7. Cacchiani V, Toth P (2012) Nominal and robust train timetabling problems. Euro J Oper Res 219(3):727–737
8. Corman F, D'Ariano A, Pacciarelli D, Pranzo M (2009) Evaluation of green wave policy in real-time railway traffic management. Transp Res Part C Emerg Technol 17(6):607–616
9. Cucala AP, Fernández A, Sicre C, Domínguez M (2012) Fuzzy optimal schedule of high speed train operation to minimize energy consumption with uncertain delays and driver's behavioral response. Eng Appl Artif Intell 25(8):1548–1557
10. Davis W (1926) The tractive resistance of electric locomotives and cars. Gen Electr Rev 29:685–707
11. Ghoseiri K, Szidarovszky F, Asgharpour MJ (2004) A multi-objective train scheduling model and solution. Transp Res Part B Methodol 38(10):927–952
12. González-Gil A, Palacin R, Batty P (2013) Sustainable urban rail systems: strategies and technologies for optimal management of regenerative braking energy. Energy Convers Manage 75:374–388

13. Goverde RMP, Hansen IA (2013) Performance indicators for railway timetables. In: 2013 IEEE international conference on intelligent rail transportation proceedings, pp 301–306
14. Goverde RMP, Corman F, D'Ariano A (2013) Railway line capacity consumption of different railway signalling systems under scheduled and disturbed conditions. J Rail Transp Plann Manage 3(3):78–94
15. Goverde RMP, Bešinović N, Binder A, Cacchiani V, Quaglietta E, Roberti R, Toth P (2016) A three-level framework for performance-based railway timetabling. Transp Res Part C Emerg Technol 67:62–83
16. Goverde RMP, Scheepmaker GM, Wang P (2021) Pseudospectral optimal train control. Euro J Oper Res 292:353–375
17. Howlett PG (2016) A new look at the rate of change of energy consumption with respect to journey time on an optimal train journey. Transp Res Part B Methodol 94:387–408
18. Howlett PG, Pudney PJ, Vu X (2009) Local energy minimization in optimal train control. Automatica 45(11):2692–2698
19. Huang K, Liao F, Gao Z (2021) An integrated model of energy-efficient timetabling of the urban rail transit system with multiple interconnected lines. Transp Res Part C Emerg Technol 129:103171
20. Liu RR, Golovitcher IM (2003) Energy-efficient operation of rail vehicles. Transp Res Part A Policy Pract 37(10):917–932
21. Lusby RM, Larsen J, Ehrgott M, Ryan D (2011) Railway track allocation: models and methods. OR Spectrum 33(4):843–883
22. Pachl J (2009) Railway operation and control. VTD Rail Publishing, Mountlake Terrace, WA, USA
23. Pachl J (2014) Timetable design principles. In: Hansen IA, Pachl J (eds) Railway timetabling and operations. Eurailpress, Hamburg, Germany, pp 13–46
24. Pontryagin LS, Boltyanskii VG, Gamkrelidze RV, Mishchenko EF (1962) The mathematical theory of optimal processes. Wiley, Hoboken, NY, USA
25. Quaglietta E, Pellegrini P, Goverde RMP, Albrecht T, Jaekel B, Marlière G, Rodriguez J, Dollevoet T, Ambrogio B, Carcasole D (2016) The ON-TIME real-time railway traffic management framework: a proof-of-concept using a scalable standardised data communication architecture. Transp Res Part C Emerg Technol 63:23–50
26. Scheepmaker GM, Goverde RMP (2015) The interplay between energy-efficient train control and scheduled running time supplements. J Rail Transp Plann Manage 5(4):225–239
27. Scheepmaker GM, Goverde RMP (2020) Energy-efficient train control using nonlinear bounded regenerative braking. Transp Res Part C Emerg Technol 121:102852
28. Scheepmaker GM, Goverde RMP (2021) Energy-efficient train timetabling considering capacity consumption and robustness. Euro J Transp Infrastr Res 21(4):1–42
29. Scheepmaker GM, Goverde RMP, Kroon LG (2017) Review of energy-efficient train control and timetabling. Euro J Oper Res 257(2):355–376
30. Scheepmaker GM, Pudney PJ, Albrecht AR, Goverde RMP, Howlett PG (2020) Optimal running time supplement distribution in train schedules for energy-efficient train control. J Rail Transp Plann Manage 14:100180
31. Sicre C, Cucala AP, Fernández-Cardador A, Jiménez JA, Ribera I, Serrano A (2010) A method to optimise train energy consumption combining manual energy efficient driving and scheduling. WIT Trans Built Environ 114:549–560
32. Su S, Li L, Tang T, Gao Z (2013) A subway train timetable optimization approach based on energy-efficient operation strategy. IEEE Trans Intell Transp Syst 14(2):883–893
33. Su S, Tang T, Li X, Gao Z (2014) Optimization of multitrain operations in a subway system. IEEE Trans Intell Transp Syst 15(2):673–684
34. Wang P, Goverde RMP (2016) Multiple-phase train trajectory optimization with signalling and operational constraints. Transp Res Part C Emerg Technol 69:255–275
35. Wang P, Goverde RMP (2017) Multi-train trajectory optimization for energy efficiency and delay recovery on single-track railway lines. Transp Res Part B Methodol 105:340–361

36. Wang P, Goverde RMP (2019) Multi-train trajectory optimization for energy-efficient timetabling. Euro J Oper Res 272(2):621–635
37. Wang Y, De Schutter B, Van den Boom T, Ning B (2013) Optimal trajectory planning for trains—a pseudospectral method and a mixed integer programming approach. Transp Res Part C 29:97–114
38. Wu C, Lu S, Xue F, Jiang L, Chen M, Yang J (2021) A two-step method for energy-efficient train operation, timetabling, and onboard energy storage device management. IEEE Trans Transp Electrif 7(3):1822–1833
39. Wu X, Dong H, Tse CK (2021) Multi-objective timetabling optimization for a two-way metro line under dynamic passenger demand. IEEE Trans Intell Transp Syst 22(8):4853–4863
40. Xu Y, Jia B, Li X, Li M, Ghiasi A (2020) An integrated micro-macro approach for high-speed railway energy-efficient timetabling problem. Transp Res Part C Emerg Technol 112:88–115
41. Yang X, Wu J, Sun H, Gao Z, Yin H, Qu Y (2019) Performance improvement of energy consumption, passenger time and robustness in metro systems: a multi-objective timetable optimization approach. Comput Industr Eng 137:106076
42. Ye H, Liu R (2016) A multiphase optimal control method for multi-train control and scheduling on railway lines. Transp Res Part B Methodol 93:377–393
43. Zhou L, Tong LC, Chen J, Tang J, Zhou X (2017) Joint optimization of high-speed train timetables and speed profiles: a unified modeling approach using space-time-speed grid networks. Transp Res Part B Methodol 97:157–181

Chapter 5
Optimisation of Train Timetables for Regenerative Braking

Xuekai Wang and Shuai Su

5.1 Introduction of Integrated Optimisation Approach

According to the mechanism of traction power transmission in railways, the traction energy from the traction power supply system is supplied to trains through the traction network. The received energy will be used to overcome the resistance during train operation or be converted into kinetic energy of the train. When the train is braking, the kinetic energy will be transformed differently due to the different braking modes. The conventional braking mode converts the kinetic energy into heat, through the friction between the track and the brake shoe [6]. In this way, all the kinetic energy is lost. In the regenerative braking mode, the motor is converted into a generator during braking, which converts the kinetic energy into regenerative braking energy (RBE) while providing the braking force. Meanwhile, the RBE that can be recycled accounts for a large proportion of the kinetic energy (e.g., about 80% for urban rail transit). Therefore, in addition to reducing the energy consumption required by train traction, making full use of RBE is another important means of energy-saving train operation [7].

RBE can be recycled in two main ways: (1) fed back to the third rail or the overhead line, and real-time reused by other trains around, and (2) use an energy storage device to store the RBE, which will be used by trains at a later time. The second approach will be detailed in the next chapter. This chapter mainly introduces the energy-saving method in the first recycling mode. When the RBE is generated by the braking trains and there are accelerating trains around, the RBE can be transmitted from the braking trains to the accelerating trains. Then, the accelerating trains use the received RBE for traction. In this way, the produced RBE is efficiently reused.

X. Wang (✉) · S. Su
Beijing Jiaotong University, Beijing, China
e-mail: 19111067@bjtu.edu.cn

S. Su
e-mail: shuaisu@bjtu.edu.cn

S. Su et al., *Energy-Efficient Train Operation*, Lecture Notes in Mobility,
https://doi.org/10.1007/978-3-031-34656-9_5

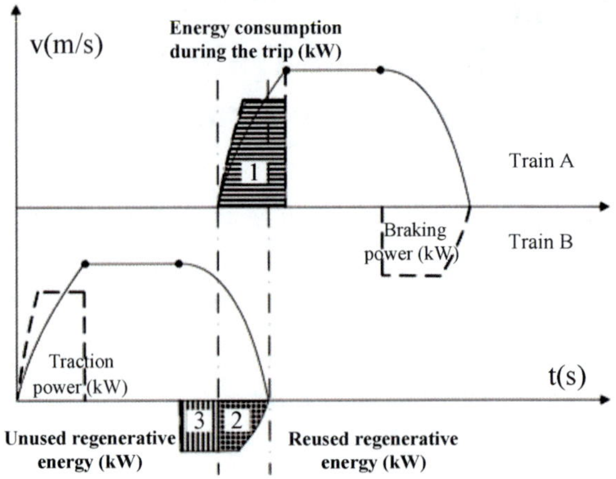

Fig. 5.1 Coordination between trains at stations

As shown in Fig. 5.1, train A departs from a station and takes the traction regime to achieve a high travel speed. Area "1" denotes the energy consumption during the accelerating process. Regenerative braking will be enabled if there is another train (Train B) braking near the same station. Then, a part of the RBE (Area "2") can be reused by the accelerating train. The other RBE (Area "3") will be consumed by the braking rheostat because there is no traction train around. In other words, the energy for towing the train is mainly provided by the traction substation while some energy may come from the regenerative braking of other trains [8].

As introduced above, the way to make full use of the RBE is to match the time that adjacent trains are applying traction and braking. Therefore, it is necessary to coordinate the operation of multiple trains. During the operation, the driving regime of trains is influenced by both the driving strategy and the train timetable. Specifically, the driving strategy determines the driving regimes applied in every moment during the inter-station operation. The train timetable determines when the train departs from the station and starts the operation between stations. Therefore, the integrated optimisation approach plays an important role to efficiently reuse the RBE. In this approach, the driving strategy and the train timetable can be simultaneously optimised, such that trains cooperate to apply traction/braking regimes.

This chapter will discuss some important contributions in the field of integrated optimisation in the metro system. Some basic assumptions are followed in this chapter, i.e., homogeneous trains with the same characteristics and stop patterns back and forth over a single metro line with a fixed speed limit throughout.

The structure of this chapter is as follows. Section 5.2 describes the mathematical models for calculating the consumed traction energy and the reused RBE. Section 5.3 introduces a timetabling optimisation approach to optimise the departure times of acceleration trains based on the arrival times of regenerative braking trains, to reuse the RBE efficiently. Section 5.4 introduces an approach which can further match the driving regimes at any time during train operation. Section 5.5 concludes with some perspectives on further developments.

5.2 Calculation of Traction Energy and Regenerative Braking Energy

5.2.1 Traction Energy Calculation Model

Given a fixed running time between stations, there are a variety of feasible driving strategies for trains to arrive at the next station on time. Normally, the traction energy consumption of different driving strategy is different [19]. As a result, this model is formulated to calculate the traction energy consumption according to the speed profile between adjacent stations. The calculation is based on a metro line as shown in Fig. 5.2. In the line, both the stations and sections are numbered as $1, 2, ..., 2L$, while station 1 and $L+1$ are two turn-back areas.

In this chapter, it is assumed that the energy consumption caused by the regenerative braking is not taken into consideration, because this part of energy is small compared to the electricity used for traction. Then, the traction energy consumption in a segment is calculated by the following formula [16].

$$E_{i,l,c}^{\text{traction}} = \int_{t_{i,l,c}^{\text{depart}}}^{t_{i,l+1,c}^{\text{arrival}}} v_i(t)u_i^+(t)\mathrm{d}t, \forall i \in I, l \in \mathcal{L} \setminus 2L, c \in \mathcal{C} \setminus C. \tag{5.1}$$

where $\mathcal{I}$, $\mathcal{L}$, $\mathcal{C}$ are the sets of trains, stations and cycles, respectively. $E_{i,l,c}^{\text{traction}}$ is the traction energy consumption between station l and $l+1$ in cycle c. All trains are assumed to be operated on the line for C cycles. Each cycle contains the circulation of each train $i \in I$ from station 1 via $L+1$ back to station 1. In each cycle, trains can adopt different driving strategies and different arrival/departure time, so as to achieve the best energy-saving effect. $t_{i,l,c}^{\text{arrival}}$ and $t_{i,l,c}^{\text{depart}}$ are the arrival time and the departure time of train i at station l in cycle c. $u_i^+(t) = \max(u_i(t), 0)$ is the traction force applied by the motor to the train. $u_i(\tau)$ is the applied force which needs to be satisfy (5.2).

$$F^{\min}(v_i(t)) \le u_i(t) \le F^{\max}(v_i(t)), \forall i \in \mathcal{I}, \tag{5.2}$$

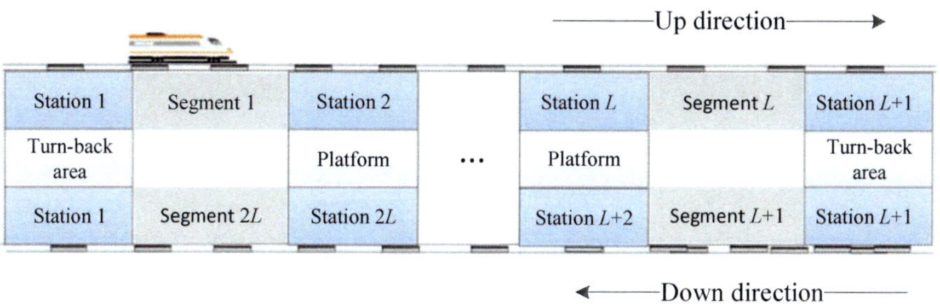

Fig. 5.2 Bi-directional urban rail transit line

where $F^{\min}$ and $F^{\max}$ are the minimum and maximum force, respectively. $v_i(t)$ is the speed of train i at t. Based on Newton's law, $v_i(t)$ can be calculated by

$$\frac{\mathrm{d}v_i}{\mathrm{d}t} = \frac{u_i(t) - q^{\mathrm{run}}(v_i(t)) - q^{\mathrm{line}}(x_i(t))}{m\vartheta}, \forall i \in \mathcal{I}. \tag{5.3}$$

In Formula (5.3), m is the train mass, ϑ is the rotating mass factor. $q^{\mathrm{run}}(v_i(t))$ is the running resistance which includes roll resistance and air resistance. The running resistance can be calculated by the following Davis Formula.

$$q^{\mathrm{run}}(v_i(t)) = a_1 v_i(t)^2 + a_2 v_i(t) + a_3, \tag{5.4}$$

where a_1, a_2 and a_3 are constant coefficients which are determined by the friction between the wheel, the track, and the aerodynamic of trains.

In Formula (5.3), $q^{\mathrm{line}}(x_i(t))$ is the line resistance mainly caused by the track gradient. The line resistance can be calculated by

$$q^{\mathrm{line}}(x_i(t)) = \mathrm{mg} \sin \beta(x_i(t)), \tag{5.5}$$

where $\beta(x_i(t))$ is the angle of gradient. $x_i(t)$ is the distance over the up direction and back in the down direction, which can be obtained by

$$\frac{\mathrm{d}x_i}{\mathrm{d}t} = v_i(t), \forall i \in \mathcal{I}. \tag{5.6}$$

Besides, the train speed should satisfy the speed limit constraint in (5.7).

$$0 \le v_i(t) \le v^{\max}(x_i(t)), \forall i \in \mathcal{I}, \tag{5.7}$$

where $v^{\max}(x_i(t))$ is the maximum train speed at position $x_i(t)$, which is formulated by

$$v^{\max}(x_i(t)) = \begin{cases} 0, \text{ if } x_i(t) \in \mathcal{Z} \\ \overline{v}, \text{ otherwise,} \end{cases} \tag{5.8}$$

where $\mathcal{Z}$ is the set of station positions. $\overline{v}$ is the speed limit in sections. Here, it is assumed that there is a constant speed limit over the entire line.

Finally, the total traction energy consumption of the whole line during the train operation is

$$E^{\mathrm{traction}} = \sum_{c=1}^{|C|} \sum_{l=1}^{2|L|-1} \sum_{i=1}^{|I|} E_{i,l,c}^{\mathrm{traction}}. \tag{5.9}$$

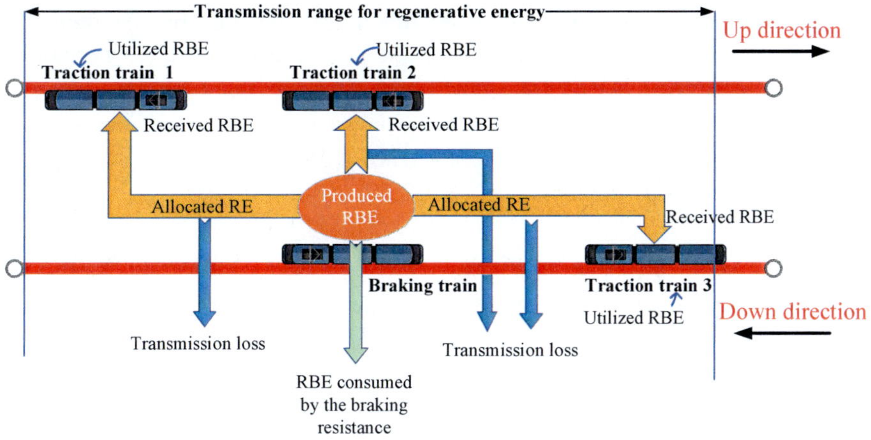

Fig. 5.3 Schematic diagram of RBE transmission

5.2.2 *Regenerative Braking Energy Calculation Model*

This model is developed to calculate the reused Regenerative Braking Energy (RBE) when multiple trains are operated in the metro line [16]. As shown in Fig. 5.3, during the train operation, the fed-back RBE can change the voltage of catenary at different locations, i.e., the voltage around the braking train is higher than the voltage around the traction train. The voltage difference makes the produced RBE be immediately transmitted to the traction trains and reused. The amount of the RBE received by each traction train is relevant to the transmission distance and the voltage difference in the traction power grid. Generally, a larger transmission distance or a lower voltage difference corresponds to a lower efficiency to reuse RBE. What is more, the RBE generated by braking trains will make the voltage of the power supply network remain at a high level if there is no traction train around or the RBE is not fully used. Therefore, the remaining RBE will be dismissed as heat, stored in ESSs, reused by the nearby accelerating trains. In this chapter, the optimisation approaches are developed based on the method in which the nearby accelerating trains reuse the produced RBE.

For train j, the amount of produced regenerative braking power can be calculated by

$$\frac{\mathrm{d}E_j^{\text{braking}}}{\mathrm{d}t} = -\varphi v_j(t) u_i^-(t), \quad \forall j \in \mathcal{I}, \tag{5.10}$$

where E_j^{braking} is the RBE produced by train j. $u_i^-(t)$ is the maximum braking force which can be calculated by $u_i^-(t) = \min(u(t), 0)$. φ is the energy conversion rate from mechanical energy to RBE. Because RBE will be lost in the power grid during transmission and the loss is huge in the long distance transmission, it is assumed that the RBE can only be transmitted to the traction trains around the braking train. The maximum transmission distance is $d^{\max}$. Then, $\mu_{i,j}(t)$ represents a binary variable to judge whether there is RBE transmitted from train j to train i at t.

$$\mu_{i,j}(t) = \begin{cases} 1, & \text{if } i \neq j, u_i(t) > 0, u_j(t) < 0, d_{i,j}(t) \leq d^{\max} \\ 0, & \text{otherwise}, \end{cases} \tag{5.11}$$

where $d_{i,j}(t)$ is the distance between trains i and j which can be obtained by

$$d_{i,j}(t) = \Big| \left|Y - \left(x_j(t) \bmod 2Y\right)\right| - \left|Y - (x_i(t) \bmod 2Y)\right| \Big|, \quad \forall i, j \in \mathcal{I}, \tag{5.12}$$

where Y is the length of the urban rail line. "mod" means the modular arithmetic. For instance, the remainder after $x_i(t)$ divided by $2Y$ is represented by "$x_i(t) \bmod 2Y$". $x_i(t)$ is the location of train i in the line. $x_i(t) <= Y$ for upstream trains, $x_i(t) > Y$ for downstream trains.

Based on the feeding circuit model, the amount of reused RBE can be precisely calculated by the load flow calculation [11, 12]. However, this method is hard to calculate the amount of reused RBE in a short time because of the computing complexity. As a result, a simplified model to approximately calculate the amount of reused RBE is applied in this chapter. In this model, the amount of reused RBE by each traction train is influenced by the transmission distance of RBE and the voltage difference in the traction power grid. The transmission distance of RBE is assumed to be represented by the actual distance between two trains, because normally the transmission distance is longer when the distance between two trains is farther. Under this assumption, the transmission distance can be represented by $d_{i,j}(t)$ which is calculated in Formula (5.12).

When more traction energy is required by the traction train or more RBE is generated by the braking train, the voltage different between two train will be more. As a result, the voltage difference is assumed to be represented by the difference of the traction power and the regenerative power. The calculation of the power difference $P_{i,j}(t)$ is given as follows.

$$P_{i,j}(t) = \max\left\{0, \frac{\mathrm{d}E_i^{\text{traction}} - \mathrm{d}E_j^{\text{braking}}}{\mathrm{d}t}\right\}, \quad \forall i, j \in \mathcal{I}, \tag{5.13}$$

where E_i^{traction} is the traction energy consumption which is calculated in Sect. 5.2.1.

To measure the volume of regenerative power that is allocated to each train, the weighting factor $\varrho_{i,j}(t)$ is defined according to the distance and the power difference [16].

$$\varrho_{i,j}(t) = \begin{cases} \varepsilon - \dfrac{d_{i,j}(t)}{x_0} + \dfrac{P_{i,j}(t)}{P_0}, & \text{if } \mu_{i,j}(t) = 1 \\ 0, & \text{if } \mu_{i,j}(t) = 0. \end{cases} \tag{5.14}$$

In Formula (5.14), ε, x_0 and P_0 are constants to guarantee that $\varrho_{i,j}(t)$ is non-negative.
Then, the RBE allocated from j to i (i.e., $E_{i,j}^{\text{allocate}}$) can be obtained by

$$\frac{\mathrm{d}E_{i,j}^{\text{allocate}}}{\mathrm{d}t} = \frac{\varrho_{i,j}(t)}{\sum_{n=1}^{|I|} \varrho_{j,n}(t)} \frac{\mathrm{d}E_{j}^{\text{braking}}}{\mathrm{d}t} \tag{5.15}$$

Some energy will be consumed by resistance during the transmission. In this approach, the simplification is made to regard the power loss to be proportional to the transmission distance and can be calculated by

$$\frac{\mathrm{d}E_{i,j}^{\text{loss}}}{\mathrm{d}t} = \theta d_{i,j}(t), \quad \forall i, j \in \mathcal{I}. \tag{5.16}$$

where $E_{i,j}^{\text{loss}}$ is the RBE loss. In terms of transmission loss, $E_{i,j}^{\text{receive}}$ is the RBE actually received by i from j, which can be obtained by

$$\frac{\mathrm{d}E_{i,j}^{\text{receive}}}{\mathrm{d}t} = \frac{\mathrm{d}E_{i,j}^{\text{allocate}} - \mathrm{d}E_{i,j}^{\text{loss}}}{\mathrm{d}t}, \quad \forall i, j \in \mathcal{I}. \tag{5.17}$$

Because the amount of reused regenerative power for each train cannot be more than the required traction power, the regenerative braking power actually reused by train i can be calculated by

$$\frac{\mathrm{d}E_{i}^{\text{regenerative}}}{\mathrm{d}t} = \min\left\{ \sum_{j=1}^{I} \frac{\mathrm{d}E_{i,j}^{\text{receive}}}{\mathrm{d}t}, \frac{\mathrm{d}E_{i}^{\text{traction}}}{\mathrm{d}t} \right\} \tag{5.18}$$

Then, the RBE reused by train i between station l and $l+1$ can be formulated by

$$E_{i,l,c}^{\text{regenerative}} = \int_{t_{i,l,c}^{\text{depart}}}^{t_{i,l+1,c}^{\text{arrival}}} E_{i}^{\text{regenerative}}(t)\mathrm{d}t \tag{5.19}$$

Finally, the total reused RBE of the whole line during the train operation is

$$E^{\text{regenerative}} = \sum_{c=1}^{|C|} \sum_{l=1}^{2L|-1} \sum_{i=1}^{|I|} E_{i,l,c}^{\text{regenerative}}. \tag{5.20}$$

5.3 Coordinated Control of Departure Times

In this chapter, an train timetable optimisation approach which can achieve the coordinated control of departure times will be introduced. The principle of the coordinated control is based on a common scenario as shown in Fig. 5.1. In the scenario, one train is going to depart from the station, when another approaching train is applying the regenerative braking in the same power supply segment [5, 17]. Then, the RBE produced by the braking train will be reused by the traction train, so the net energy consumption (i.e., the difference between the traction energy and the reused RBE) can be reduced. Based on this idea, a timetabling approach by optimising the departure times is discussed om this chapter.

5.3.1 *Solution Approach*

The formulation of the optimisation problem is divided into two parts, i.e., the optimal train control model and the coordinated model. In the optimal train control model, the optimal driving strategy between stations and the traction energy consumption during the train operation are calculated. Based on the solution of the optimal train control model, the coordinated model is then used to obtain the optimal departure time of the train to minimise the net energy consumption.

Optimal Train Control Model

The method to calculate the energy-efficient driving strategy given a fixed inter-station running time will be introduced in this section. The similar content is also considered in Chap. 3 (for one segment) and Chap. 4 (successive segments over a line). Compared to them, a difference optimisation method will be introduced in this chapter considering a closed (up and down) line.

Firstly, based on the traction energy calculation model introduced in Sect. 5.2, the optimal train control model between stations is formulated as

$$\begin{cases} \min\, E = \displaystyle\int_0^T u^+(t)v(t)\mathrm{d}t \\ s.t.\ m\vartheta\dfrac{\mathrm{d}v}{\mathrm{d}t} = u(t) - q^{\text{run}}(v) - q^{\text{line}}(x), \\ \qquad \dfrac{\mathrm{d}x}{\mathrm{d}t} = v(t), \\ \qquad v_0 = v_T = 0, \quad 0 \le v \le v^{\max}, \\ \qquad -f^{\max} \le u \le F^{\max}, \end{cases} \tag{5.21}$$

where T is a variable which represents the inter-station running time. During the train operation, the inter-station running time for each segment is normally different. For simplicity, we suppress the dependency on the specific segment here.

Table 5.1 Optimal driving regimes

u_o	Condition	Regime
$u^{\max}$	$p_2 > v$	Maximum acceleration (MA)
$[0, u^{\max}]$	$p_2 = v$	Cruising with partial power (CR)
0	$0 < p_2 < v$	Coasting (CO)
$[u^{\min}, 0]$	$p_2 = 0$	Cruising with partial braking (CR)
$u^{\min}$	$p_2 < 0$	Maximum braking (MB)

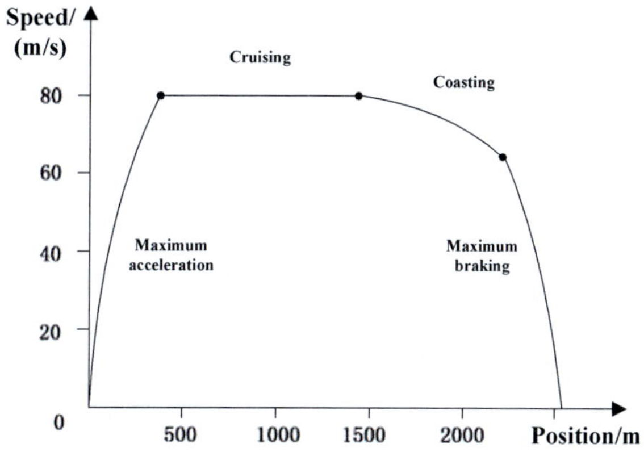

Fig. 5.4 Optimaldriving strategy on a section with constant gradient and speed limit

By applying the Pontryagin maximum principle, the optimal control sequences should maximize the Hamiltonian function H with respect to the control variable u,

$$H = p_1 v(t) + p_2(u(t) - q^{\text{run}}(v) - q^{\text{line}}(x)) - u^+(t)v(t). \tag{5.22}$$

Therefore, there are five optimal driving regimes as shown in Table 5.1.

As introduced in Chap. 3, the optimal driving strategy should be designed as follows: after departing from a station, the maximum acceleration must be applied. If the optimal cruising speed exceeds the speed limit and the maximum speed is reached, then applying the cruising regime. Then, depending on the available running time reserve, the coasting phase should be then used. Finally, the train takes the maximum braking regime to stop at the next station (see Fig. 5.4). On steep descents, coasting may precede cruising at the speed limit.

For a segment with constant gradient and speed limit, the minimum energy consumption is uniquely determined by the trip time and vice versa (see Fig. 5.5). Hence, the optimal driving strategy can be calculated with either the known trip time or the known energy consumption [9].

In this chapter, an iterated method to get the optimal driving strategy based on the given energy consumption will be introduced. In each iteration, a traction energy consumption required for the inter-station train operation will be given according to the method introduced later. Then the shortest inter-station running time under

Fig. 5.5 Decreasing character of the E-T function

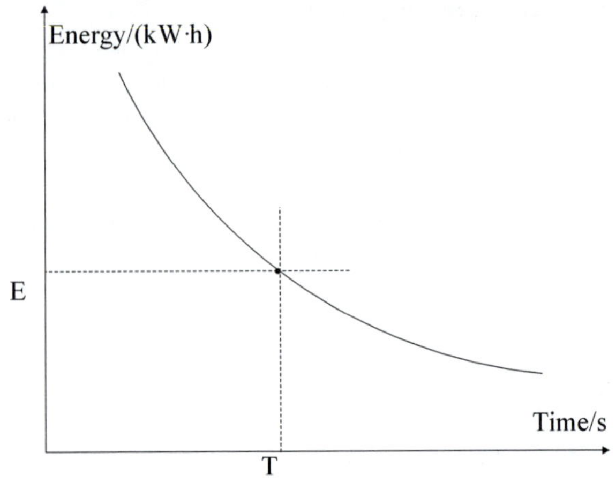

the given energy consumption will be calculated. If the calculated running time is approximately equal to the specified inter-station running time, the iteration will be stopped; otherwise, the given traction energy for train operation will be changed for the next iteration.

When the energy consumption between stations is given, the shortest running time can be calculated by the following method. Firstly, divide the segment into several units according to the distance. The gradient and speed limit in each unit are constants. The acceleration is assumed to be constant at each distance unit. Then, the speed sequences in the MA regime are calculated with the given energy consumption. The speed sequence represents the set of train speed at each instant during one regime. If the train speed reaches the speed limit during the acceleration process, the rest energy will be used for generating the speed sequences of the CR phase. The rest of train journey consists a CO speed sequence $\{v_k^{j'}\}$ and a MB $\{v_b^{j''}\}$. $\{v_k^{j'}\}$ begins from the end speed of the CR regime and $\{v_b^{j''}\}$ ends at the final speed v_T. To obtain the switching point between CO and MB, the envelope curve of CO is calculated from the end time of CR, while the envelope curve of MB is calculated from T forward. Then, the intersection of two envelope curves is the switching point. The optimal speed of CO and MB regimes are the minimum speed value of the two envelope curves.

The details for obtaining the speed trajectory of a given segment under a given traction energy is described in Algorithm 1. In the algorithm, the differential equations in (5.21) is discretized into the difference equations according to the distance. Δx is a small distance unit, e.g., $\Delta x = 1$ m.

After obtaining the speed sequence for the given traction energy generated by Algorithm 1, the energy unit (a small amount of energy which is defined as 0.03 kWh here) will be attempted to distribute to different sections for achieving the corresponding time reductions. After a comparison among these time reductions, the energy unit will be finally distributed to the section that can achieve the maximum

Algorithm 1: Optimal speed sequences for a given traction energy in the segment

Step 1: Initialize the initial speed v_0^j and the energy consumption E_j for section j;

Step 2: Divide the section into n_j pieces such that the distance of each piece is Δx. It is assumed that the length of each section is a multiple of Δx;

Step 3: Generate the speed sequences for the MA phase, where $F(v_i^j)$ is the maximum traction force;
$v_i^j = v_0^j$, while $E_j > 0$, $v_i^j < V$, do
$(v_{i+1}^j)^2 - (v_i^j)^2 = 2\Delta x(F(v_i^j)/m\vartheta - q^{\text{run}}(v_i^j)/m\vartheta - q^{\text{line}}(x_i))/m\vartheta$,
$E = E - F(v_i^j)\Delta x$,
$i = i + 1$;

Step 4: If the speed v_i^j reaches the speed limit and there is still remaining energy ($E > 0$), go to Step 5; If there is no remaining energy ($E = 0$) before the train reaches the speed limit, go to Step 6,

Step 5: Under the assumption that the optimal cruising speed is the speed limit, when the speed v_i^j reaches the speed limit, then partial braking or partial power is applied to keep cruising and the speed sequences are calculated as $v_{i+1}^j = v_i^j$, go to Step 6

Step 6: Generate the speed envelope curve for CO phase. Set $k = i$, $v_b^{j'} = v_i^j$.
While $k \le n_j$, do
$(v_{k+1}^{j'})^2 - (v_k^{j'})^2 = 2\Delta x(-q^{\text{run}}(v_k^{j'})/m\vartheta - q^{\text{line}}(x_k)/m\vartheta)$, $k = k + 1$;

Step 7: Generate the speed envelope curve for MB phase. Set $b = n_j$, $v_b^{j''} = v_i^j$.
While $b > i$, do
$(v_{b+1}^{j''})^2 - (v_b^{j''})^2 = 2\Delta x(-B(v_b^{j''})/m\vartheta - r(v_b^{j''})/m\vartheta - g(x_b)/m\vartheta)$, $b = b - 1$;

Step 8: Get the optimal speed of CO and MB regimes according to two envelope curves.
While $i \le n_j$, do
$v_i^j = \min(v_k^{j'}, v_b^{j''})$, $i = i + 1$;

Step 9: Return the optimal speed sequences and the trip time of this segment

$$T_j = \sum_{i=0}^{n_j} \frac{\Delta x}{v_i^j}.$$

time reduction. This distribution process will be repeated to shorten the primary trip time until the practical trip time is delivered, after which the driving strategy will also be obtained as well. The flow chart of the algorithm is described in Fig. 5.6.

The Cooperative Model

When considering the transmission of RBE between one accelerating and one braking train, two important elements of the net energy consumption are the traction energy consumption during the trip and the reused amount of the regenerative braking energy. The former element is related to the trip time and driving strategy between stations and the later element depends on the coordinated control between the accelerating and braking trains. In order to satisfy the requirement of punctuality, accelerating trains need to arrive at the next station on time. Hence, a delayed departure time implies a shorter running time between stations, which could increase the energy consumption during the journey (see Fig. 5.5). Besides, the energy consumption between stations

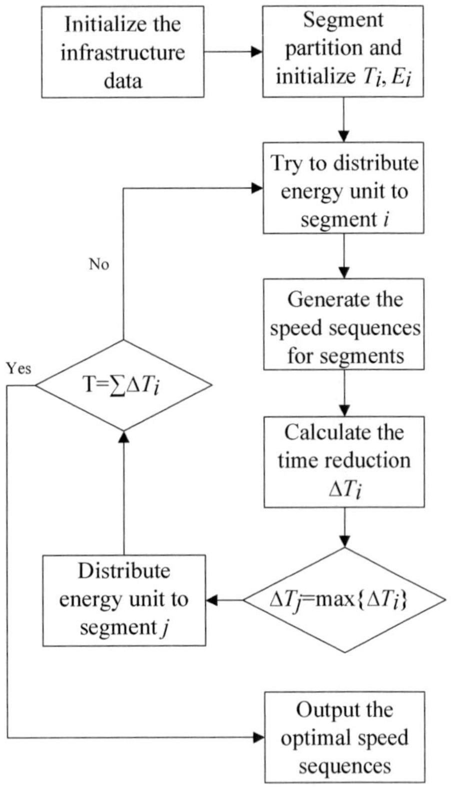

Fig. 5.6 The flow chart for calculating the optimal driving strategy for trains between stations

increases when the train has a departure delay. However, a later departure may create an opportunity for making better use of the RBE. Consequently, we formulate a coordinated model to deal with this trade-off problem to minimise the practical energy consumption by solving the optimal departure time for the accelerating train.

For the calculation of the practical energy consumption, we firstly obtain the speed sequences of the braking train as well as the available regenerative energy at each time slice. With a given departure time, we then can calculate the speed sequence of the departure train and obtain the energy consumption needed for each time slice during the acceleration process as well as the total energy consumption E for the trip. Finally, the reused regenerative energy of each time slice can be calculated and summed to be E_r according to

$$E_r = \int_{t=0}^{T} \min\{E^a(\overline{s}, t), E^b(\overline{s}, t)\}\mathrm{d}t, \tag{5.23}$$

where $E^a(\overline{s}, t)$ and $E^b(\overline{s}, t)$ are the potential use of the regenerative energy for the accelerating trains and the available RBE from the braking trains at the time t, respec-

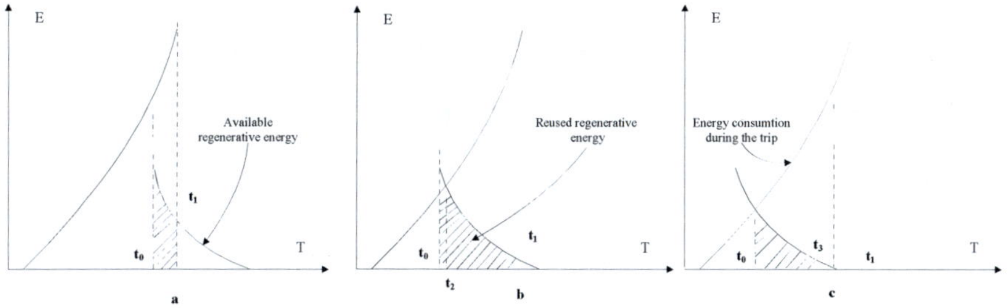

t_0—time that trains start to brake t_1—time that trains stop accelerating t_3—time that trains stop braking

Fig. 5.7 Coordination between trains at stations

tively. $\overline{s}$ means that the positions of the cooperative trains are close enough to utilize the RBE, not implying that trains are in the same position. When the cooperative operation between trains is enabled, both $E_a(\overline{s}, t)$ and $E_b(\overline{s}, t)$ are positive. If not, $E_b(\overline{s}, t)$ will be zero.

During the coordination between trains to reuse RBE, the distance between the accelerating and braking trains should not be too large, so as to avoid the loss of RBE during transmission. It is noted that the accelerating distance of the metro vehicles is approximately 200 m, which is near the substation. Furthermore, the accelerating processes for trains are similar to each other. Hence, the average energy losses due to the energy transfer between trains and substations are assumed to be constant. Additionally, the cooperative control of two trains at stations happens between close trains and the energy losses due to the energy transfer between trains are assumed to be fixed.

Specifically, Fig. 5.7 shows the specific coordinated processes for different departure times with the given braking speed sequences. According to the traction characteristic, when the acceleration is constant, the train with faster speed needs higher energy consumption. As a result, the required traction energy increases with respect to T. On the contrary, trains have a large amount of kinetic energy at high speed and less kinetic energy at low speed. Thus, the available regenerative energy is a decreasing function.

For the coordinated state (a), $E^a(\overline{s}, t) > E^b(\overline{s}, t)$. According to Equation (5.18),

$$E_r = \int_{t=0}^{T} \min\{E^a(\overline{s}, t), E^b(\overline{s}, t)\}\mathrm{d}t = \int_{t_0}^{t_1} E^b(\overline{s}, t)\mathrm{d}t. \tag{5.24}$$

where t_0 is a constant which represents the time that starts to apply the braking force. t_1 is the time that the trains stop accelerating, which is determined by the departure time. E_r will be an increasing function with respect to t_1 during process (a). Since $E^b(\overline{s}, t) > 0$ during the interval $[t_0, t_1]$. t_0 and t_1 are the beginning time to apply the regenerative braking and the end time to apply the traction force.

For the process (b), E_r can be calculated as

$$E_r = \int_{t_0}^{t_2} E^a(\overline{s}, t)\mathrm{d}t + \int_{t_2}^{t_1} E^b(\overline{s}, t)\mathrm{d}t, \tag{5.25}$$

from which we can obtain

$$\frac{\mathrm{d}E_r}{\mathrm{d}t_2} = E^a(\overline{s}, t_2) + \frac{\mathrm{d}t_1}{\mathrm{d}t_2} E^b(\overline{s}, t_1) - E^b(\overline{s}, t_2) = \frac{\mathrm{d}t_1}{\mathrm{d}t_2} E^b(\overline{s}, t_1). \tag{5.26}$$

where t_2 is the time point when the traction and braking power are equal.

To obtain the relationship between t_1 and t_2, it is assumed that t_1 and t_2 are not fixed firstly. Then, as the departure time of the acceleration train is delayed, the white area (energy consumption curve) in process (b) will move to the right. In this way, t_2 will increase and likewise t_2 from the point that two curves intersect. Thus, t_1 will increase with t_2, $\mathrm{d}t_1/\mathrm{d}t_2 > 0$. Besides, $E^b(\overline{s}, t_1)$ is positive. Hence, $\mathrm{d}E_r/\mathrm{d}t_2 > 0$ and the amount of reused regenerative energy will keep increasing.

When the coordinated process reaches the state (c), E_r should be presented as the following equation.

$$E_r = \int_{t_0}^{t_3} E^b(\overline{s}, t)\mathrm{d}t, \tag{5.27}$$

where E_r is a decrease function with respect to t_3. In other words, the amount of regenerative energy will decrease with a later departure time in this situation. t_3 is the end time of the regenerative braking.

Hence, the amount of reused regenerative energy shows an increasing, and then decreasing trend from the state (a)–(c). Based on the continuity of the function of E_r with respect to departure time, we conclude that the amount of reused regenerative energy is a unimodal function with respect to t_1, i.e.,

$$\begin{cases} \dfrac{\mathrm{d}E_r}{\mathrm{d}t_1} > 0, & \text{when } \ t < t^*; \\ \dfrac{\mathrm{d}E_r}{\mathrm{d}t_1} = 0, & \text{when } \ t = t^*; \\ \dfrac{\mathrm{d}E_r}{\mathrm{d}t_1} < 0, & \text{when } \ t > t^*, \end{cases} \tag{5.28}$$

where t^* is the time when the reused RBE peaks.

In addition, E increases with the a later departure time,

$$\frac{\mathrm{d}E}{\mathrm{d}t_1} > 0, \tag{5.29}$$

and

$$\frac{\mathrm{d}E_p}{\mathrm{d}t_1} = \frac{\mathrm{d}E}{\mathrm{d}t_1} - \frac{\mathrm{d}E_r}{\mathrm{d}t_1}, \tag{5.30}$$

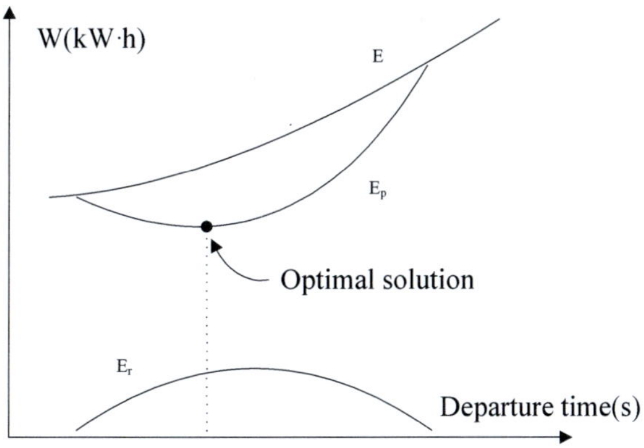

Fig. 5.8 Optimal departure time

where E_p is the net energy consumption. From (5.28) (5.29), and (5.30), we analyze the function E_p as the following two cases.

- If $\mathrm{d}E/\mathrm{d}t_1 > \mathrm{d}E_r/\mathrm{d}t_1$, i.e., $\mathrm{d}E_p/\mathrm{d}t_1 > 0$, it implies that it will cost more practical energy consumption with a later departure time. In other words, trains should leave the station as early as possible to reduce the practical energy consumption.
- If $\mathrm{d}E/\mathrm{d}t_1 < \mathrm{d}E_r/\mathrm{d}t_1$, we can obtain $\mathrm{d}E_p/\mathrm{d}t_1 < 0$, which implies that it will cost less practical energy consumption with a later departure time.

In conclusion, the traction energy consumption between stations increases and the reused RBE decreases when the train has a departure delay (see Fig. 5.8). there must be one inflexion at most for the practical energy consumption function. Hence, we solve the optimal departure time by using the bisection method, which is shown in Fig. 5.9.

First, the earliest departure time x^0 and latest departure time x^q are initialized. The gradient of the practical energy consumption for these two departure times are then calculated. If both of the gradients are positive, it implies that the practical energy consumption will increase with a later departure time and the earliest departure time is the optimal solution. Otherwise, the extremum will be obtained by shrinking the range of the best solution with bisection method until the defined procedure is completed.

In practice, the proposed method can be off-line or real-time applied in the metro systems.

In the off-line mode, the proposed method can further optimise the departure time of each train in the existing train timetable, to get a better energy-saving effect. In addition, the computation time of this approach is short enough (average 0.4 s) to be applied to the real-time train timetable adjustment in the metro system. Nowadays, many metro systems are equipped with the Communication Based Train Control

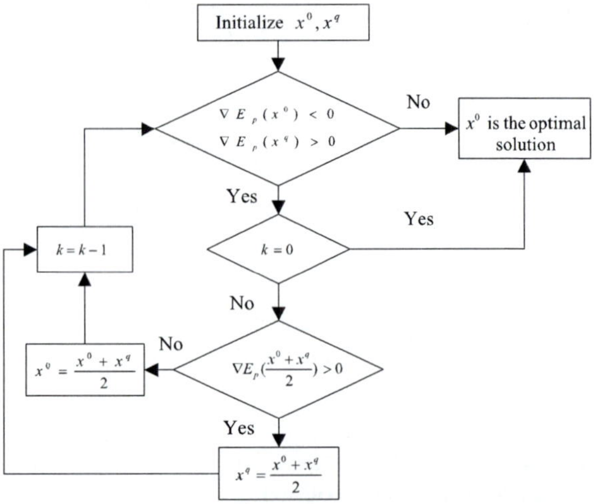

Fig. 5.9 Flow chart of bisection method applied to $\nabla E_p(x)$

(CBTC) system, which can implement communication among trains, track equipment, and the control center. The train position and speed are known more accurate than the traditional signalling system, which lays the foundation for cooperation among trains and makes it possible to efficiently manage the railway traffic. The sequence diagram of the cooperation process in subway systems is described in Fig. 5.10. Firstly, sensors on trains obtain the train information (such as train position, and train speed), and Vehicle On-Board Controllers (VOBCs) send this information to a Zone Controller (ZC) and Operation Control Center (OCC) through the Data Communication System (DCS), after which a ZC will generate the moving authority for braking trains. Then, based on the possible departure time, the OCC calculates the optimal departure time to achieve a good cooperation between trains for the departure train. Finally, the OCC feeds back the departure information to trains through the DCS and the VOBCs on the trains can calculate the driving strategy for the following journey with the given moving authority.

5.3.2 Examples

Two examples based on the Beijing Metro Yizhuang line are presented in this section to show the effectiveness of the integrated optimisation approach.

Example A

The first example presents a coordinated control for two trains running between Jiugong and Yizhuangqiao station.

Train A has stopped in the Jiugong station to allow passengers to alight and board. The original arrival time is $27'35$ and the earliest departure time is $27'58$

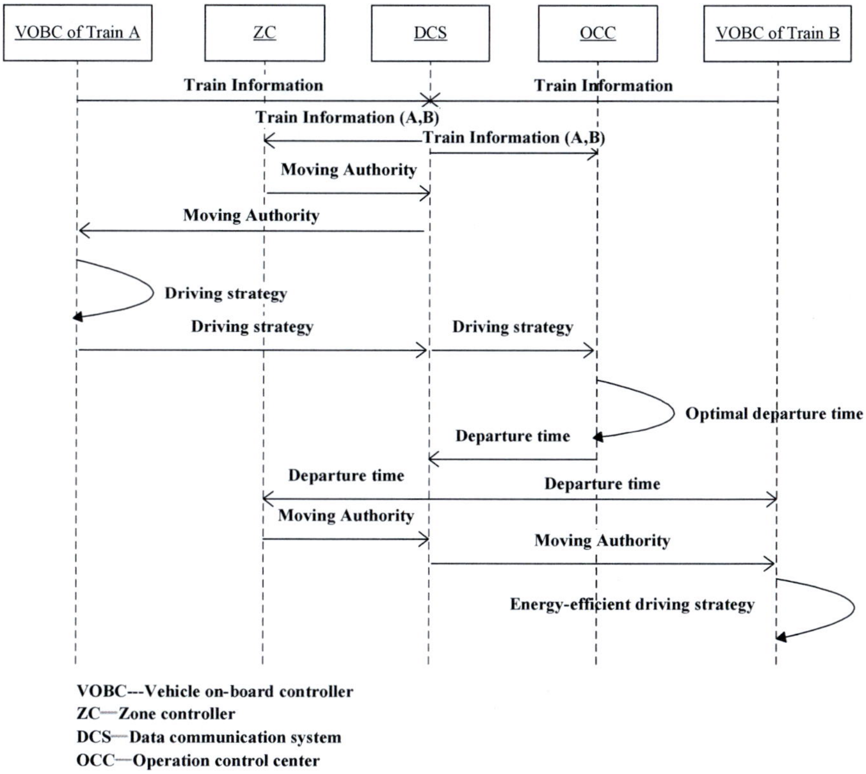

Fig. 5.10 The sequence diagram of the cooperation process

in off-peak hours (The shortest dwell time mainly includes the door-open time, passengers' boarding and alighting time, and the door-closing time). In addition, the latest departure time is 28′20 and the train should arrive in the Yizhuangqiao station at 30′30 to follow the punctuality according to the planned timetable. Train B is running in the segment between Yizhuangqiao and Jiugong and starts to brake at 28′05 for stopping in the Jiugong station at the opposite track

By applying the proposed algorithm, the optimal departure time is solved as 28′07, and the convergence process of the algorithm is presented in Fig. 5.11.

The details of the energy consumption for different departure schedules is also presented in Table 5.2. Without considering the cooperation between trains, the departing train follows the earliest departure time and the energy consumption calculated from the optimal driving strategy is 13.600 kWh. No regenerative energy is reused. Hence, the practical energy consumption without changing the departure time is 13.600 kWh. However, by adjusting the departure time with the coordinated model, the energy consumption during the trip is calculated to be 14.229 kWh, and the regenerative energy is 0.956 kWh. Hence, the practical energy consumption calculated from the coordinated model is 13.273 kWh (see Table 5.2), which reduces

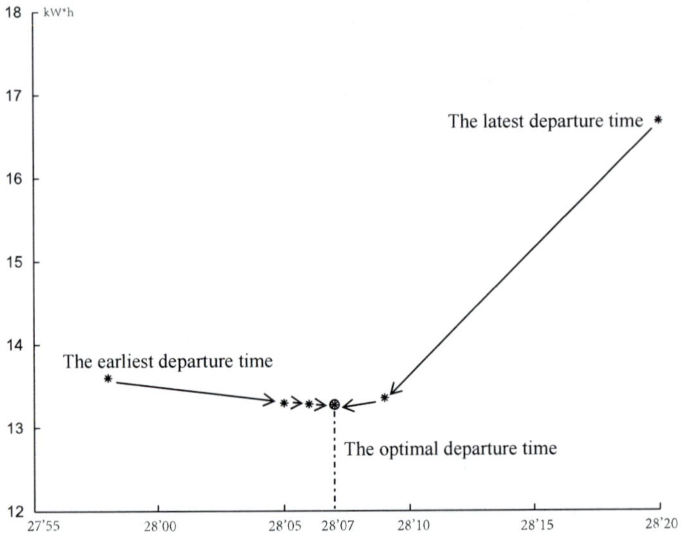

Fig. 5.11 Convergence of the bisection method for the off-peak hours example

Table 5.2 Comparison on the energy consumptions for different departure time

Departure time	T(s)	E(kWh)	E_r(kWh)	E_p(kWh)
27′58	152	13.600	0	13.600
28′00	150	13.795	0.207	13.588
28′03	147	14.002	0.615	13.387
28′05	145	14.143	0.848	13.295
28′07	143	14.229	0.956	13.273
28′08	142	14.343	1.044	13.299
28′09	141	14.471	1.116	13.355
28′10	140	14.700	1.160	13.540
28′11	139	14.943	1.192	13.751
28′13	137	15.414	1.192	14.222
28′16	134	16.129	1.092	15.037
28′20	130	17.414	0.736	16.678

the practical energy consumption by 2.4%. In addition, the optimal driving strategy obtained from the coordinated model is shown in Fig. 5.12.

There is another scenario that the train arrives at the same station in peak hours or a delay happens, resulting in that the earliest departure time becomes 28′08, while the state of the braking train does not change. Then the coordinated approach is applied to solve the optimal departure time, which is shown in Fig. 5.13. The result implies that the earlier the train departs from the station, the less energy consumption it will cost. Since the accelerating train has missed the best opportunity to cooperate with

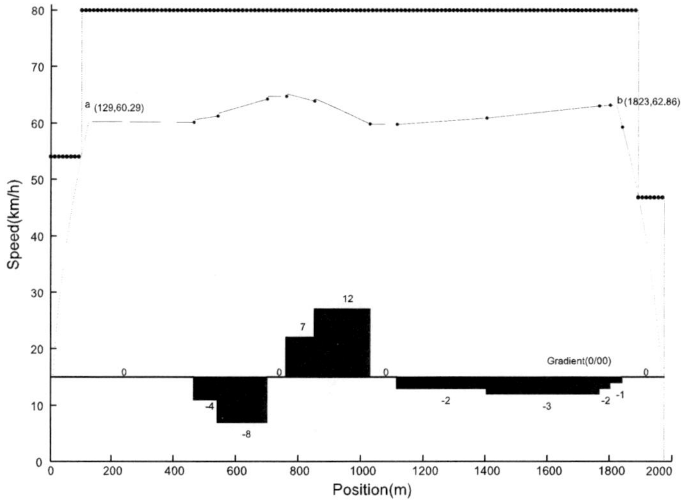

Fig. 5.12 Optimal driving strategy for the departure train

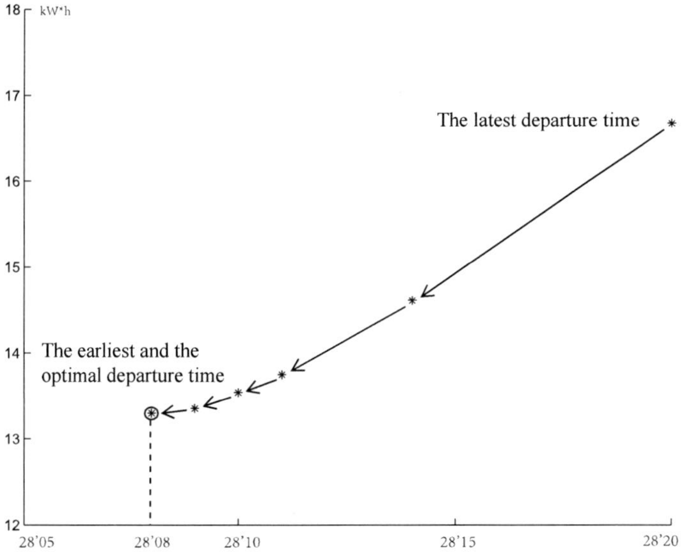

Fig. 5.13 Convergence of the bisection method for the peak hours example

the braking train and the practical energy consumption mainly depends on the energy consumption during the trip.

Example B

To efficiently utilize the RBE, the proposed approach achieves the coordination between trains by changing the departure time, which is scheduled by the timetable.

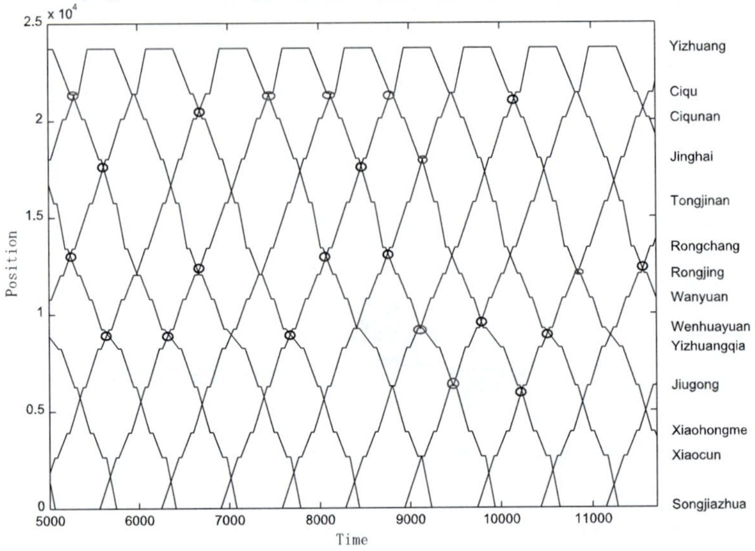

Fig. 5.14 Modification of timetable for trains coordination, the circles denote the generated coordinated operation for trains

Hence, in Example B, we apply the proposed approach to modify the timetable for a fixed time period 8:00–9:00 AM. This time period is one of the peak hours with more passengers alighting and boarding the Yizhuang line. Generally, eleven trains will be put into service to meet the passenger demand with an approximately cyclic timetable. The arrival time of each train is fixed in this case. As shown in Table 5.3, the energy consumption in practice for the 8:00–9:00 AM period (case 1) is 3451.43 kWh, in which the regenerative energy accounts for 38.90 kWh. Then we apply the optimal train control approach (without coordination) to optimise the driving strategy for trains during the time period 8:00–9:00 AM and the practical energy consumption can be reduced by 10.87%, because the energy consumption during the trip is efficiently reduced. It should be noted that a small difference occurs to the amount of regenerative energy due to the change of control sequences when the optimal driving strategy is applied (see the case 2 in Table 5.3). However, the optimal control model does not contribute to efficiently use the regenerative energy.

Finally, we apply the coordinated approach for the same time period and calculate the corresponding energy consumption (case 3 in Table 5.3). For the coordinated approach, the energy consumption shows a gentle rise since some trip times are shortened, resulting in more energy. A great increase in the utilization of the regenerative energy is achieved from 40.41 kWh to 86.73 kWh. Specifically, many acceleration-braking train pairs are efficiently coordinated, which is shown in Fig. 5.14.

Hence, the practical energy consumption is ultimately reduced by 11.34% and the proposed coordinated approach provides a better performance on energy-saving, which achieves more energy reduction by 0.47% than the optimal train control model.

Table 5.3 Comparison on energy consumption during 8:00–9:00 AM

8:00–9:00 AM	E (kWh)	E_r (kWh)	E_p (kWh)	R (%)
Case 1	3451.43	38.90	3412.53	–
Case 2	3081.86	40.41	3041.45	10.87
Case 3	3112.41	86.73	3025.68	11.34

The calculations were performed on a laptop with processor speed of 2.4 GHz and memory size of 3 GB. The average computation time of the proposed algorithm is about 0.4 s, which can meet the requirement of real-time control.

5.4 Integrated Schedule and Train Trajectory Optimisation for Metro Lines

The integrated optimisation approach introduced in the last chapter reused the produced RBE efficiently by synchronising the departure time to the arrival time of trains. Apart from the arrival time and departure time, the driving strategy between stations also influences the reuse of RBE [4, 15, 18]. For example, the sequence of the driving regimes can be optimised to increase the synchronization times of the traction/braking regimes and increase the amount of reused RBE. As a result, it is necessary to study the synchronous optimisation method, in which the driving strategy and train timetable are jointly optimised. Then, the net energy consumption of train operation can be systematically reduced [10]. In this chapter, an synchronous optimisation method based on the Space-Time-Speed (STS) network methodology is introduced. In the STS network, the three-dimensional train trajectories can depict the driving strategy and the train timetable simultaneously. As a result, the impacts of both the driving strategy and train timetable on the reuse of RBE can be jointly considered with the coordinated control. A set of mathematic models are presented in Sect. 5.4.1 to formulate the integrated optimisation problem for metro lines. In Sect. 5.4.2, a Multi-Agent Reinforcement Learning (MARL) algorithm is designed to obtain the optimised train timetable and driving strategy. In Sect. 5.4.3, a set of examples based on an urban rail transit line is given.

5.4.1 Mathematical Formulation of Integrated Optimisation

Space-Time-Speed network model

In this part, the integrated optimisation problem is transformed into a multi-step decision problem based on the Space-Time-Speed (STS) network methodology.

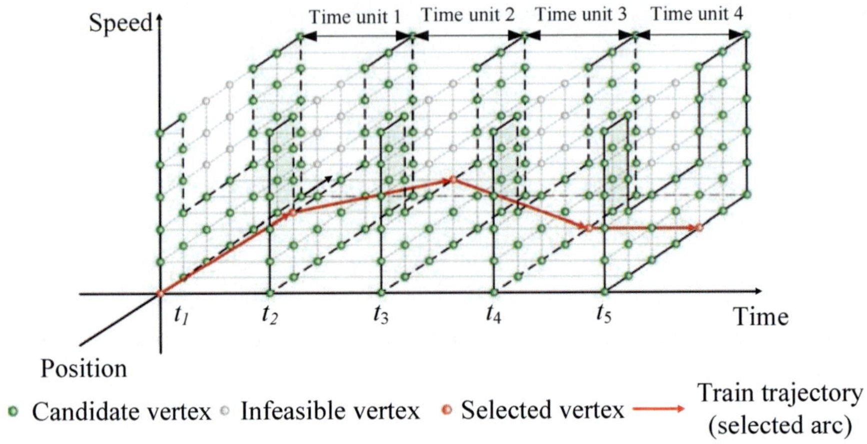

Fig. 5.15 STS network

Firstly, the running time, speed and position of trains are discretized to generate a set of vertexes in the three-dimensional space (i.e., STS network) as shown in Fig. 5.15 [14]. The total running time is divided into $K - 1$ time units according to the interval Δt. The initial time of the kth time unit is $t_k = k\Delta t$. Besides, the terminal time of train operation (i.e., the terminal time of the time unit K) is $K\Delta t$. $x_{i,k}$ and $v_{i,k}$ are the position and speed of train i at t_k, respectively. $d_{i,k}$ is the running distance of train i until the time point t_k. Then, the index of running cycle for train i at t_k is expressed as Eq. 5.31. The running cycle represents the journey between two consecutive departures from the original station.

$$c_{i,k} = \lceil \frac{x_{i,1} + d_{i,k}}{2Y} \rceil, \quad \forall i \in \mathcal{I}, k \in \mathcal{K}, \tag{5.31}$$

where Y is the length of metro line. $x_{i,k}$ can be obtained by the following equation.

$$x_{i,k} = x_{i,1} + d_{i,k} - 2Y(c_{i,k} - 1), \quad \forall i \in \mathcal{I}, k \in \mathcal{K} \setminus 1. \tag{5.32}$$

Each vertex in the STS networks represents a train state, and each arc connecting the adjacent vertexes represents a three-dimensional train trajectory in the time unit. The projection of the three dimensional train trajectory in the speed-time describes the control characteristics, while the projection in the space-time side depicts the timetable characteristics. By selecting a vertex for each time point and connecting adjacent vertexes with an arc in the chronological order, the train trajectory for each train can be obtained. By projecting the train trajectory onto a plane, the driving strategy of each train and the train timetable can be obtained.

When the neighbouring states in the STS network are given, the driving strategy between the states can be determined according to the principle as shown in Table 5.4.

Table 5.4 Principle to determine the driving strategy between states

Condition	Sequence of driving regimes
$v_{i,k} = 0, v_{i,k+1} = 0, x_{i,k} = x_{i,k+1}$	DW
$v_{i,k+1} = v_{i,k+1}^{\max}, x_{i,k+1} = x_{i,k+1}^{\max}$	MA
$v_{i,k} \neq 0, v_{i,k+1}^{\text{coast}} < v_{i,k+1} < v_{i,k+1}^{\max}, x_{i,k+1}^{\text{coast}} < x_{i,k+1} < x_{i,k+1}^{\max}$	PA-CO
$v_{i,k} = 0, v_{i,k+1} < v_{i,k+1}^{\max}, x_{i,k+1} < x_{i,k+1}^{\max}$	DW-PA
$v_{i,k+1} = v_{i,k+1}^{\text{coast}}, x_{i,k+1} = x_{i,k+1}^{\text{coast}}$	CO
$v_{i,k+1} \neq 0, v_{i,k+1}^{\min} < v_{i,k+1} < v_{i,k+1}^{\text{coast}}, x_{i,k+1}^{\min} < x_{i,k+1} < x_{i,k+1}^{\text{coast}}$	CO-PB
$v_{i,k+1} = 0, x_{i,k+1} < x_{i,k+1}^{\min}$	PB-DW
$v_{i,k+1} = v_{i,k+1}^{\min}, x_{i,k+1} = x_{i,k+1}^{\min}$	MB

In Table 5.4, There are six possible driving regimes which include Dwell (DW), Partial Acceleration (PA), Partial Braking (PB), Coasting (CO), Maximum Acceleration (MA), Maximum Braking (MB). Moreover, $v_{i,k+1}^{\max}$, $v_{i,k+1}^{\text{coast}}$ and $v_{i,k+1}^{\min}$ are the train speed at t_{k+1} when the MA, CO and MB are adopted in the time unit k, respectively. Similarly, $x_{i,k+1}^{\max}$, $x_{i,k+1}^{\text{coast}}$ and $x_{i,k+1}^{\min}$ are the position of train at t_{k+1} when applying MA, CO and MB.

Energy calculation model

According to the calculation method in Sect. 5.2, the traction energy and the reused RBE are calculated in this model. After the discretization by time, the total traction energy consumption can be calculated by summing up the traction energy in each time unit. The traction energy of train i between time points t_k and t_{k+1} can be calculated by the following equation.

$$E_{i,k,k+1}^{\text{traction}} = \int_{t_k}^{t_{k+1}} v_i(\tau)u_i^+(\tau)\mathrm{d}\tau, \forall i \in I, k \in \mathcal{K} \setminus K. \tag{5.33}$$

The train speed $v_i(\tau)$ and the train position $s_i(\tau)$ at time τ can be calculated by (5.34) and (5.35), respectively. Equations (5.2) and (5.7) should also be met during train operation.

$$v_i(\tau) = \int_{t_1}^{\tau} \frac{u_i(\tau) - q^{\text{run}}(v_i(\tau)) - q^{\text{line}}(x_i(\tau))}{m}\mathrm{d}\tau, \forall i \in \mathcal{I}, \tau \in [0, t_K]. \tag{5.34}$$

$$x_i(\tau) = x_{i,1} + \int_{t_1}^{\tau} v_i(\tau)\mathrm{d}\tau, \forall i \in \mathcal{I}, \tau \in [0, t_K]. \tag{5.35}$$

By the discretization of three-dimensional variables in (5.10)–(5.19), the reused RBE in $[t_k, t_{k+1}]$ can be calculated by the following equations.

$$E^{\text{braking}}_{j,k,k+1} = \int_{t_k}^{t_{k+1}} \varphi v_j(\tau) u_i^-(\tau) \mathrm{d}\tau, \forall j \in \mathcal{I}, k \in \mathcal{K} \setminus K, \tag{5.36}$$

$$E^{\text{allocate}}_{i,ji,k,k+1} = E^{\text{braking}}_{j,k,k+1} \frac{\varpi_{i,j,k,k+1}}{\sum_{\nu=1}^{|I|} \varpi_{j,\nu,k,k+1}}, \forall i, j \in \mathcal{I}, k \in \mathcal{K} \setminus K, \tag{5.37}$$

$$\varpi_{i,j,k,k+1} = \begin{cases} \varrho - \dfrac{d_{i,j,k}}{x_0} + \dfrac{E_{i,j,k,k+1}}{E_0}, & \text{if } \zeta_{i,j,k,k+1} = 1 \\ 0, & \text{if } \zeta_{i,j,k,k+1} = 0 \end{cases} \tag{5.38}$$
$$\forall i, j \in \mathcal{I}, k \in \mathcal{K} \setminus K,$$

$$d_{i,j,k} = \Big| \left|Y - (x_{j,k} \bmod 2Y)\right| - \left|Y - (x_{i,k} \bmod 2Y)\right| \Big|, \forall i, j \in \mathcal{I}, k \in \mathcal{K}. \tag{5.39}$$

$$E_{i,j,k,k+1} = E^{\text{traction}}_{i,k,k+1} - E^{\text{braking}}_{j,k,k+1}, \forall i, j \in \mathcal{I}, k \in \mathcal{K} \setminus K, \tag{5.40}$$

$$\mu_{i,j,k,k+1} = \begin{cases} 1, & \text{if } i \neq j, u_i(t_k) > 0, u_j(t_k) < 0, x_{i,j,k} \leq d^{\max} \\ 0, & \text{otherwise} \end{cases} \tag{5.41}$$
$$\forall i, j \in \mathcal{I}, k \in \mathcal{K} \setminus K,$$

$$E^{\text{loss}}_{i,j,k,k+1} = \theta d_{i,j,k}, \forall i, j \in \mathcal{I}, k \in \mathcal{K} \setminus K, \tag{5.42}$$

$$E^{\text{receive}}_{i,j,k,k+1} = \begin{cases} E^{\text{allocate}}_{i,j,k,k+1} - E^{\text{loss}}_{i,j,k,k+1}, & \text{if } \mu_{i,j,t_k} = 1 \\ 0, & \text{if } \mu_{i,j,k,k+1} = 0 \end{cases} \forall i, j \in \mathcal{I}, k \in \mathcal{K} \setminus K. \tag{5.43}$$

$$E^{\text{regenerative}}_{i,k,k+1} = \max\{E^{\text{traction}}_{i,k,k+1}, \sum_{\nu=1}^{|I|} E^{\text{receive}}_{i,j,k,k+1}\}, \forall i \in \mathcal{I}, k \in \mathcal{K} \setminus K. \tag{5.44}$$

Train operation model

To ensure the safety of train operation and meet the passenger demand, the constraints for the train operation are given in this model.

Firstly, the headway constraint is formulated as Formula (5.45) to ensure the operation safety.

$$t_{i+1,l,c}^{\text{depart}} - t_{i,l,c}^{\text{depart}} \geq h^{\min}, \forall i \in \mathcal{I} \setminus I, k \in \mathcal{K}, \tag{5.45}$$

where $h^{\min}$ is the minimum headway. $t_{i,l,c}^{\text{depart}}$ is the time when the train i departs from station l in cycle c, which can be obtained by

$$t_{i,l,c}^{\text{depart}} = t_k, \text{ if } v_{i,k} = 0, v_{i,k+1} > 0, x_{i,k} = p_l, c_{i,k} = c, \quad \forall i \in \mathcal{I}, l \in \mathcal{L}, c \in \mathcal{C} \tag{5.46}$$

Then, the dwell time should meet (5.47).

$$t_l^{\text{dwell,min}} \leq t_{i,l,c}^{\text{depart}} - t_{i,l,c}^{\text{arrive}} \leq t_l^{\text{dwell,max}}, \forall i \in \mathcal{I}, l \in \mathcal{L}, c \in \mathcal{C}, \tag{5.47}$$

where $t_l^{\text{dwell,min}}$ and $t_l^{\text{dwell,max}}$ are the lower bound and the upper bound of the dwell time at station l, respectively. $t_l^{\text{dwell,min}}$ and $t_l^{\text{dwell,max}}$ are determined by the opening time and closing time of the doors, the amount of alighting and boarding passengers, the loading and unloading speed and the width of door, etc. $t_{i,l,c}^{\text{arrive}}$ is the arrival time of the train i at station l in cycle c, which is obtained by

$$t_{i,l,c}^{\text{arrive}} = t_k, \text{ if } v_{i,k-1} > 0, v_{i,k} = 0, x_{i,k} = p_l, c_{i,k} = c, \quad \forall i \in \mathcal{I}, l \in \mathcal{L}, c \in \mathcal{C}. \tag{5.48}$$

Meanwhile, the constraints of the inter-station running time are presented as Formulas (5.49) and (5.50).

$$t_l^{\text{trip,min}} \leq t_{i,l+1,c}^{\text{arrive}} - t_{i,l,c}^{\text{depart}} \leq t_l^{\text{trip,max}}, \quad \forall i \in \mathcal{I}, l \in \mathcal{L} \setminus 2L, c \in \mathcal{C}. \tag{5.49}$$

$$t_l^{\text{trip,min}} \leq t_{i,1,c+1}^{\text{arrive}} - t_{i,2L,c}^{\text{depart}} \leq t_l^{\text{trip,max}}, \quad \forall i \in \mathcal{I}, c \in \mathcal{C} \setminus C. \tag{5.50}$$

Finally, the cycle time should meet

$$t^{\text{cycle,min}} \leq t_{i,1,c+1}^{\text{depart}} - t_{i,1,c}^{\text{depart}} \leq t^{\text{cycle,max}}, \forall i \in \mathcal{I}, c \in \mathcal{C} \setminus C. \tag{5.51}$$

Integrated optimisation model

In this part, the integrated optimisation model for the energy-efficient train operation is summarized . The objective function is to minimise the net energy consumption:

$$\min\ E^{\text{net}} = \sum_{k=1}^{K-1} \sum_{i=1}^{|I|} \left(E_{i,k,k+1}^{\text{traction}} - E_{i,k,k+1}^{\text{regenerative}} \right). \tag{5.52}$$

In Formula (5.52), E^{net} is the net energy consumption of the train operation. $E_{i,k,k+1}^{\text{traction}}$ and $E_{i,k,k+1}^{\text{regenerative}}$ can be attained by the traction energy calculation model and the reused RBE calculation model model, respectively.

The constraints for the cooperative control include Formulas (5.2), (5.7) (5.47) (5.49) (5.50) (5.51).

5.4.2 Solution Approach

An improved MARL algorithm is designed to solve the integrated optimisation problem. Reinforcement Learning (RL) is rapidly developing in these years to solve the sequential decision-making problems [20]. In the RL algorithm, agent will repeatedly make decisions, interact with an environment, and update the policy to make better decisions. MARL is a kind of RL algorithm which can address the sequential decision-making problem of multiple autonomous agents in a common environment. In the cooperative setting, agents interact with each other and collaborate to optimise a common long-term return. Besides, MARL is a model-free control method which is suitable for solving the nonlinear control issues and the multi-variable control issues [1].

The energy-efficient model in this chapter is highly non-linear and has a large number of decision variables (i.e., the speed and the position of trains). Therefore, a cooperative MARL algorithm is adopted. Specifically, each train is considered as an agent which can observe the states of all trains by bi-directional communication. Then, a train can autonomously make a decision about what driving strategy will be adopted in the next time unit. The environment in the algorithm is the power supply system which provides the energy for train operation. The cooperative setting is adopted and the common goal for trains is minimising the net energy consumption.

Introduction of the MARL algorithm

In this section, the applied MARL algorithm is an improved algorithm based on [3].

In the MARL algorithm, the framework is collaborative multi-agent Markov Decision Process (MDP), which can be defined by a tuple $\langle \mathcal{S}, \mathcal{A}, \mathcal{R}, \gamma \rangle$. $\mathcal{S} = \mathcal{S}_1 \times \ldots \times \mathcal{S}_I$ is the joint state space of agents. The joint action space is represented by $\mathcal{A} = \mathcal{A}_1 \times \ldots \times \mathcal{A}_I$. The state transition function is $\mathcal{P} = \mathcal{S} * \mathcal{A} * \mathcal{S} \rightarrow [0, 1]$ which gives the transition probability $p(\mathbf{s}'|\mathbf{s}_k, \mathbf{a}_k)$ that the system will move to the state $\mathbf{s}'$ after executing the action $\mathbf{a}_k$ at $\mathbf{s}_k$. After taking the action, agents can obtain the reward $R_1 = \ldots = R_I$ according to a reward function $\mathcal{S} \times \mathcal{A} \rightarrow \mathcal{R}$. In the cooperative setting, the reward obtained by each agent is the same. Therefore, R is used to represent the common reward of agents.

In the algorithm, the deterministic policy $\pi(\mathbf{s}_k)$ is adopted, which is a mapping from the joint state $\mathbf{s}_k$ to a deterministic joint action $\mathbf{a}_k$. The MARL algorithm aims to find the optimal policy $\pi^*(\mathbf{s}_k)$ to maximize the expected discounted return. One efficient way to get the optimal policy is to estimate the optimal Q-function. The Q-learning algorithm is an on-policy Temporal Difference (TD) control algorithm which is widely used to estimate the optimal Q-function. In the algorithm, the estimated Q-functions are stored in a tabulation. Agents can repeatedly employ the ε-greedy approach to select the action based on the estimated Q-function, and update the estimated Q-function according to the following formula [13].

$$Q^{n+1}(\mathbf{s}_k, \mathbf{a}_k) = Q^n(\mathbf{s}_k, \mathbf{a}_k) + \alpha[R(\mathbf{s}_k, \mathbf{a}_k) + \gamma \max_{\mathbf{a}'} Q^n(\mathbf{s}_{k+1}, \mathbf{a}')], \tag{5.53}$$

where $Q^n(\mathbf{s}_k, \mathbf{a}_k)$ is the estimated Q-function in the nth iteration. By repeatedly executing the action and updating the estimated Q-function, the optimal Q-function can be obtained by iteration.

However, the space of the joint state and the joint action exponentially grows when the number of the agents increases, which makes a high memory requirement for the problem with many agents when the Q-learning algorithm is applied. Thus, the MARL algorithm is used to reduce the memory requirement for the multi-agent problem. Value Function Factorization (VFF) method is the key idea of the applied MARL algorithm, where the global Q-function is decomposed into several local Q-functions according to the Coordination Graph (CG) $\mathcal{G} = (\mathcal{V}, \mathcal{E})$. $\mathcal{V}$ is the vertex set and $\mathcal{E}$ is the edge set. Each vertex represents an agent. The connected agents indicate a local coordination dependency. In this way, a set of local Q-functions is produced. The edge $(i, j) \in \mathcal{E}$ and the local Q-function $Q_{(i,j)}$ is one-one corresponded, which means that the number of local Q-functions is equal to the edge number. As shown in the following equation, the value of the global Q-function is the sum of all local Q-functions. Therefore, the optimal global Q-function can be obtained by estimating each local Q-function.

$$Q^n(\mathbf{s}_k, \mathbf{a}_k) = \sum_{(i,j)\in\mathcal{E}} Q^n_{(i,j)}(\mathbf{s}_{(i,j),k}, a_{i,k}, a_{j,k}). \tag{5.54}$$

To estimate the optimal local Q-functions, agents should repeatedly execute the chosen action and update the estimated local Q-functions according to the observation. With probability ϵ, agents will choose a random joint action $\mathbf{a}_k$, otherwise will choose the action according to

$$\mathbf{a}_k = [a_{1,k}, \ldots, a_{I,k}] = \arg\max_{a_{1,k},\ldots,a_{I,k}} \sum_{(i,j)\in\mathcal{E}} Q^n_{(i,j)}(\mathbf{s}_{(i,j),k}, a_{i,k}, a_{j,k}). \tag{5.55}$$

Then, each local Q-function will be updated. Formula (5.53) can be transformed to the following form according to Formula (5.54),

$$
\begin{aligned}
&\frac{1}{2}\sum_{j\in\Gamma_i} Q^{n+1}_{(i,j)}(\mathbf{s}_{(i,j),k}, a_{i,k}, a_{j,k}) \\
&= \frac{1}{2}(1-\alpha)\sum_{j\in\Gamma_i} Q^{n}_{(i,j)}(\mathbf{s}_{(i,j),k}, a_{i,k}, a_{j,k}) \\
&+ \alpha\left[r_i(\mathbf{s}_k, \mathbf{a}_k) + \gamma\frac{1}{2}\sum_{j\in\Gamma_i} Q^{n}_{(i,j)}(\mathbf{s}_{(i,j),k+1}, a_i', a_j') \right],
\end{aligned}
\tag{5.56}
$$

where Γ_i is the set of agents that are connected with agent i in the CG. $r_i(\mathbf{s}_k, \mathbf{a}_k)$ represents the amount of individual reward allocated to the agent i. Recall that the reward of each agent is the same in the cooperative setting. $r_i(\mathbf{s}_k, \mathbf{a}_k)$ is only a part of $R(\mathbf{s}_k, \mathbf{a}_k)$ and is used to update the local Q-function. $r_i(\mathbf{s}_k, \mathbf{a}_k)$ measures the effect of the action performed by agent i, and the cooperation between agent i and its neighbors on the total reward $R(\mathbf{s}_k, \mathbf{a}_k)$. The relationship between $r_i(\mathbf{s}_k, \mathbf{a}_k)$ and $R(\mathbf{s}_k, \mathbf{a}_k)$ is formulated as

$$
R(\mathbf{s}_k, \mathbf{a}_k) = \sum_{i=1}^{|I|} r_i(\mathbf{s}_k, \mathbf{a}_k),
\tag{5.57}
$$

To update the local Q-function of each edge, $r_i(\mathbf{s}_k, \mathbf{a}_k)$ should be further divided and allocated to the edges connecting agent i and its neighbours $j \in \Gamma_i$. In the designed MARL algorithm, the individual reward $r_i(\mathbf{s}_k, \mathbf{a}_k)$ of agent i is divided into two parts, i.e., $r_i^{\text{in}}(\mathbf{s}_k, \mathbf{a}_k)$ and $r_i^{\text{co}}(\mathbf{s}_k, \mathbf{a}_k)$. Specifically, $r_i^{\text{in}}(\mathbf{s}_k, \mathbf{a}_k)$ represents the part of individual reward only caused by the actions taken by agent i, which means that $r_i^{\text{in}}(\mathbf{s}_k, \mathbf{a}_k)$ is not related to the action of other agents. Therefore, $r_i^{\text{in}}(\mathbf{s}_k, \mathbf{a}_k)$ can be equally divided to the edges around the agent i. $r_i^{\text{co}}(\mathbf{s}_k, \mathbf{a}_k)$ is the part of individual reward caused by the cooperation between agent i and the the neighbouring agents, which need to be distributed to edges by weight $\omega_{i,(i,j)}$. The more the impacts of the cooperation between agents i and j on the individual reward, the more the weight $\omega_{i,(i,j)}$ is. Besides, $\sum_{j\in\Gamma_i} \omega_{i,(i,j)} = 1$.

Based on this idea, the new update method of the local Q-functions can be formulated as

$$
\begin{aligned}
Q^{n+1}_{(i,j)}(\mathbf{s}_{(i,j),k}, a_{i,k}, a_{j,k}) &= (1-\alpha)Q^{n+1}_{(i,j)}(\mathbf{s}_{(i,j),k}, a_{i,k}, a_{j,k}) \\
&+ \alpha\left[\frac{r_i^{\text{in}}(\mathbf{s}_k, \mathbf{a}_k)}{\Gamma_i}\right. \\
&+ \omega_{i,(i,j)} r_i^{\text{co}}(\mathbf{s}_k, \mathbf{a}_k) + \frac{r_j^{\text{in}}(\mathbf{s}_k, \mathbf{a}_k)}{\Gamma_j} \\
&\left.+ \omega_{j,(i,j)} r_j^{\text{co}}(\mathbf{s}_k, \mathbf{a}_k) + \gamma Q^{n}_{(i,j)}(\mathbf{s}_{(i,j),k+1}, a_i', a_j')\right].
\end{aligned}
\tag{5.58}
$$

This update method can integrate the known cooperative mechanism among agents with the distribution of the individual reward. As a result, the utilization of the knowledge about the cooperative control model has the potential to obtain a better solution in shorter time.

By repeatedly choosing the action according to Formula (5.55) and updating the local Q-functions according to Formula (5.58) for $N^{\max}$ times, an approximate optimal policy can be obtained based on the estimated local Q-functions by

$$\pi^{\text{apx}}(\mathbf{s}_k) = \underset{a_{1,k},\ldots,a_{I,k}}{\arg\max}\Big(\sum_{(i,j)\in\mathcal{E}} Q_{(i,j)}^{N^{\max}}(s_{(i,j)}, a_i, a_j)\Big). \tag{5.59}$$

Framework of the MARL algorithm for integrated optimisation

In the integrated optimisation problem, each train is an agent which can observe the states of all trains by bi-directional communication and make the decision about what driving strategy will be adopted. The environment is the power supply system which provides the electrical energy for train operation. The definition of the state, action, reward, and state transition probability for the cooperative control problem will be introduced in detail.

- **State**: The joint state of agents at t_k is composed of the states of all trains $\mathbf{s}_k = [\mathbf{s}_{1,k}, \ldots, \mathbf{s}_{I,k}]$, where the state of train i is defined as $\mathbf{s}_{k,i} = (T_{i,k}, v_{i,k}, x_{i,k})$. In the vector, $v_{i,k}$ and $x_{i,k}$ are the speed and the position of train i at t_k, respectively. $T_{i,k}$ is the span of time between t_k and the latest time when train i departs from or arrives at a station. $T_{i,k}$ can be obtained by the following formula.

$$T_{i,k} = \begin{cases} t_k - t_{i,l,c_{i,k}}^{\text{arrive}}, & \text{if } x_{i,t_k} \in \mathcal{Z} \\ t_k - t_{i,l,c_{i,k}}^{\text{depart}}, & \text{otherwise} \end{cases} \quad \forall i \in \mathcal{I}, k \in \mathcal{K}. \tag{5.60}$$

  To generate the state space of trains, the variables T, v, x are firstly discreted according to the intervals Δt, Δv, and Δx, respectively. Then, the state space is shown in Fig. 5.16. Different from the STS network, the horizontal axis of the state network represents the span of time T since the latest time when the train departs from or arrives at a station, instead of the running time t since the original time point. At each time point k, the train should reach a state in Fig. 5.16. There are two kinds of states, i.e., the station state and the section state.
- **Action**: $\mathbf{a}_k = [a_{1,k}, \ldots, a_{I,k}]$ is the joint action of all trains at t_k. $a_{i,k}$ is the action of train i which represents the driving strategy adopted by train i in the time unit k. After adopting the selected driving strategy, the train will reach another state in Fig. 5.16. There are four kinds of action, i.e., dwell action, departure action, arrival action and running action.
- **Reward**: Under the cooperative setting, the common reward $R(\mathbf{s}_k, \mathbf{a}_k)$ is received by each train. To reduce the net energy consumption of all trains, $R(\mathbf{s}_k, \mathbf{a}_k)$ is defined as the negative value of the net energy consumption which can be calculated by Formula (5.61).

$$R(\mathbf{s}_k, \mathbf{a}_k) = -\sum_{i=1}^{|I|} \left(E_{i,k,k+1}^{\text{traction}} - E_{i,k,k+1}^{\text{regenerative}} \right). \tag{5.61}$$

Then, the common reward will be divided into each agent and each edge to update the local Q-functions by using the proposed MARL algorithm. Firstly, the individual reward for each agent is the net energy consumption of each train $r_i(\mathbf{s}_k, \mathbf{a}_k) = E_{i,k,k+1}^{\text{traction}} - E_{i,k,k+1}^{\text{regenerative}}$. $r_i(\mathbf{s}_k, \mathbf{a}_k)$ can be divided into independent individual reward and the cooperative individual reward according to the cooperative relationship among trains. Specifically, the independent individual reward $r_i^{\text{in}}(\mathbf{s}_k, \mathbf{a}_k)$ which is only caused by the driving strategy of train i is the traction energy consumption $E_{i,k,k+1}^{\text{traction}}$. The cooperative individual reward $r_i^{\text{co}}(\mathbf{s}_k, \mathbf{a}_k)$ represents the utilized RBE ($E_{i,k,k+1}^{\text{regenerative}}$) of train i which is caused by the synchronization of traction/braking regimes between train i and other trains.

Next, the individual reward will be allocated to each edge (i, j) according to the transmission mechanism of RBE, i.e., more cooperative individual reward will be allocated to the edge connecting the trains whose driving regimes are synchronized to reuse more RBE. In this way, the distribution weight $\omega_{i,(i,j)}$ to measure the impact of edge (i, j) on the $r_i^{\text{co}}(\mathbf{s}_k, \mathbf{a}_k)$ can be calculated by

$$\omega_{i,(i,j)} = \frac{E_{i,j,k,k+1}^{\text{allocate}}}{\sum_{v \in \Gamma_i} E_{v,i,k,k+1}^{\text{allocate}}}, \tag{5.62}$$

Finally, the reward which is allocated to edge (i, j) can be calculated by

$$r_{(i,j)}(\mathbf{s}_k, \mathbf{a}_k) = \frac{E_{i,k,k+1}^{\text{traction}} + E_{j,k,k+1}^{\text{traction}}}{|I| - 1} + \omega_{i,(i,j)} E_{i,k,k+1}^{\text{regenerative}} + \omega_{j,(i,j)} E_{j,k,k+1}^{\text{regenerative}}. \tag{5.63}$$

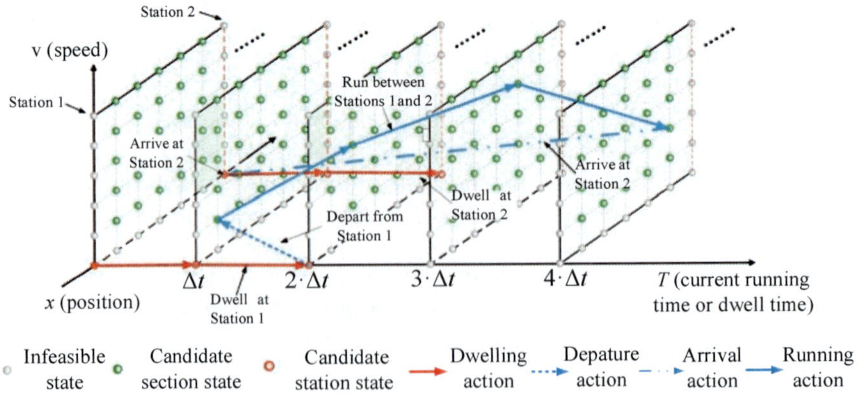

Fig. 5.16 Diagram of the state and action spaces for the cooperative control problem

MARL algorithm design for cooperative control

In this part, the procedures of the MARL algorithm for the integrated optimisation problem is developed as follows.

1. Initialize the parameters. The initial state of trains is set to be $\mathbf{s}_1 = [\mathbf{s}_{1,1}, \ldots, \mathbf{s}_{I,1}]$, where $\mathbf{s}_{i,1} = (x_{i,1}, 0, 0)$. The number of time steps K in an episode is calculated by

$$K = \frac{Ct^{\text{cycle,max}}}{\Delta t} + 1. \tag{5.64}$$

2. Generate the space of the joint state and the joint action as introduced in 5.4.2.
3. Take a sample in each time step. When the joint state of trains is $\mathbf{s}_k$, a joint action $\mathbf{a}_k$ will be chosen from the space of the feasible actions and be taken by the agents. Then, trains will reach a new joint state $\mathbf{s}_{k+1}$ and the reward will be obtained by trains from the environment. From $\mathcal{A}_k$, a joint action $\mathbf{a}_k$ will be chosen by using the ε-greedy method. With probability ε, a random joint action will be chosen. Otherwise, the joint action will be adopted according to

$$\mathbf{a}_k = [a_{1,k}, \ldots, a_{I,k}] = \arg\max_{a_{1,k},\ldots,a_{I,k}} \sum_{(i,j)\in\mathcal{E}} Q^n_{(i,j)}(\mathbf{s}_{(i,j),k}, a_{i,k}, a_{j,k}). \tag{5.65}$$

After executing the chosen driving strategies, the joint state will be deterministically transformed to be $\mathbf{s}_{i,k+1} = (T_{i,k+1}, v_{i,k+1}, x_{i,k+1})$ according to

$$v_{i,k+1} = v_{i,k} + \int_{t_k}^{t_{k+1}} \frac{u_i(\tau) - q^{\text{run}}(v_i(\tau)) - q^{\text{line}}(x_i(\tau))}{m} \mathrm{d}\tau, \quad \forall i \in \mathcal{I}, k \in \mathcal{K} \setminus K. \tag{5.66}$$

$$x_{i,k+1} = x_{i,k} + \int_{t_k}^{t_{k+1}} v_i(\tau)\mathrm{d}\tau, \forall i \in \mathcal{I}, k \in \mathcal{K} \setminus K. \tag{5.67}$$

$$T_{i,k+1} = \begin{cases} T_{i,k} + \Delta t, & \text{if } v_{i,k} > 0, v_{i,k+1} > 0 \text{ or } v_{i,k} = 0, v_{i,k+1} = 0 \\ 0, & \text{otherwise} \end{cases} \quad \forall i \in \mathcal{I}, k \in \mathcal{K} \setminus K. \tag{5.68}$$

4. Update the local Q-functions. Based on the samples, the local Q-functions can be updated by

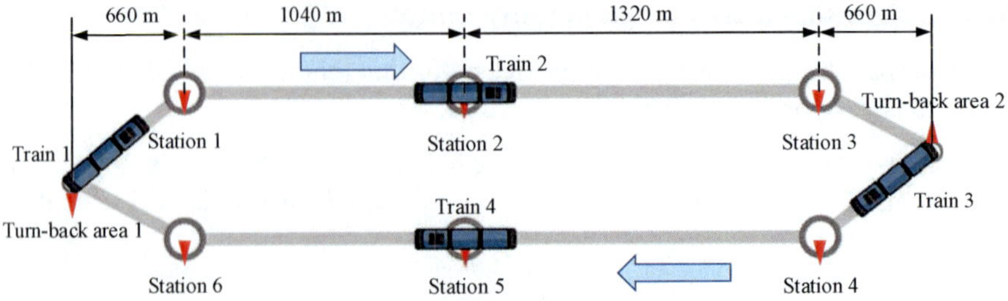

Fig. 5.17 Schematic diagram of the metro line

$$\begin{aligned} Q^{n+1}_{(i,j)}(\mathbf{s}_{(i,j),k}, a_{i,k}, a_{j,k}) &= (1-\alpha) Q^{n}_{(i,j)}(\mathbf{s}_{(i,j),k}, a_{i,k}, a_{j,k}) \\ &+ \alpha \Bigg[\frac{E^{\text{traction}}_{i,k,k+1} + E^{\text{traction}}_{j,k,k+1}}{I-1} \\ &+ \frac{E^{\text{allocate}}_{i,j,k,k+1}}{\sum_{v \in \Gamma_i} E^{\text{allocate}}_{v,i,k,k+1}} E^{\text{regenerative}}_{i,k,k+1} + \frac{E^{\text{allocate}}_{i,j,k,k+1}}{\sum_{v \in \Gamma_j} E^{\text{allocate}}_{v,j,k,k+1}} E^{\text{regenerative}}_{j,k,k+1} \\ &+ \gamma Q^{n}_{(i,j)}(\mathbf{s}_{(i,j),k+1}, a'_i, a'_j) \Bigg]. \end{aligned} \tag{5.69}$$

5. Repeatedly takes samples by step 3 and update the local Q-functions by step 4 until the iteration time reaches $N^{\max}$.
6. Stop iteration and calculate the approximate optimal policy by

$$\pi^{\text{apx}}(\mathbf{s}_k) = \arg\max_{a_{1,k},\ldots,a_{I,k}} \Big(\sum_{(i,j)\in\mathcal{E}} Q^{N^{\max}}_{(i,j)}(s_{(i,j)}, a_i, a_j) \Big). \tag{5.70}$$

5.4.3 Examples

In this section, a set of case studies is conducted based on a metro line (see Fig. 5.17) to show the effectiveness of the proposed method.

In the examples, all trains are operated on the line for one cycle. The results of MARL method are measured by three benchmarks. In the first benchmark, the driving strategy of trains are not optimised. In the second benchmark, only the driving strategy of each train is optimised to reduce the traction energy. In the third benchmark, the integrated optimisation problem is calculated by applying the Dynamic Programming (DP) algorithm [16].

The results of the proposed method and benchmarks are listed in Table 5.5. Next, the results of different methods will be compared and analyzed.

Table 5.5 Result data of the proposed method and four benchmarks for three cases

Number of trains	Optimisation category	Algorithm	Traction energy (kWh)	Reuse RBE (kW·h)	Net energy (kW·h)	Convergence time (s)
3	Without optimisation	–	222.8	22.8	200.0	–
3	Driving strategy optimisation	DP	201.8	12.6	189.2	–
3	Integrated optimisation	DP	224.4	57.2	167.2	866.3
3	Integrated optimisation	MARL	227.0	58.8	168.2	716.3
4	Without optimisation	–	297.1	29.8	267.3	–
4	Driving strategy optimisation	DP	269.2	31.1	238.1	–
4	Integrated optimisation	DP	295.5	95.0	200.5	5639.1
4	Integrated optimisation	MARL	293.8	91.3	202.5	2249.8
5	Without optimisation	–	371.5	61.2	310.3	–
5	Driving strategy optimisation	DP	336.4	48.8	287.6	–
5	Integrated optimisation	DP	403.6	169.3	234.3	754435.6
5	Integrated optimisation	MARL	407.0	165.1	242.6	15791.1

Compared to the the results without optimisation, the net energy consumption is reduced by 15.9, 24.2 and 19.5% in the three cases when the proposed method is used, respectively. On one hand, less traction energy consumption is required after optimising the driving strategy by applying the integrated optimisation. On the other hand, more RBE can be reused because the synchronization of the traction/braking regimes is considered.

Compared to the driving strategy optimisation method in the three-train circumstance, the application of integrated optimisation coupled with the MARL leads to a slight increase in the traction energy by 12.5% and a significant reduction in the reused RBE by 366.7%. This means an overall 11.1% decrease in the net energy consumption, and the figures are 15.0 and 15.6% under the four-train and five-train circumstances, respectively.

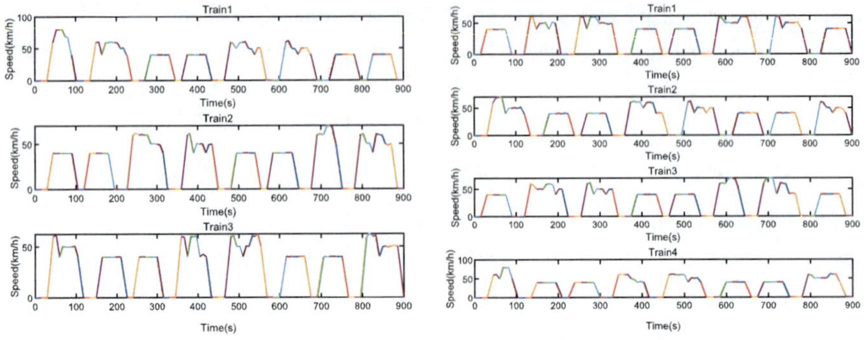

(a) Three operated trains (b) Four operated trains

Fig. 5.18 Driving strategy of trains by using the proposed method

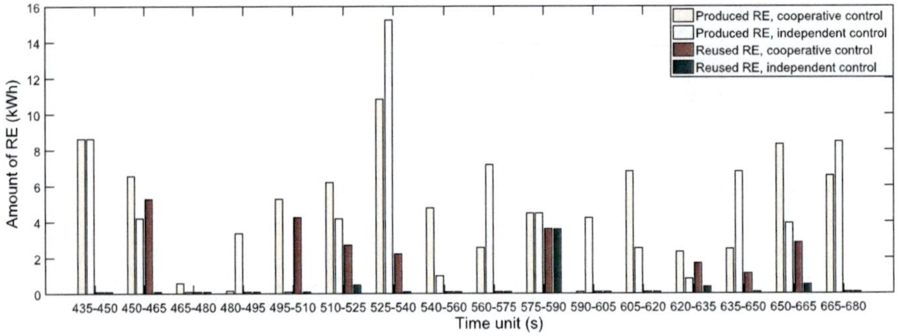

Fig. 5.19 Diagram of the produced RBE and the reused RBE in each time unit

The main reason for the reduction of the net energy consumption is the efficient reuse of RBE. In Fig. 5.18, each arc with different color represents the speed curve in a time unit. Then, the traction/braking regimes of trains can be synchronized by using the proposed method. As a result, more RBE can be reused and the net energy consumption is decreased. For example, the amount of the produced RBE and reused RBE in each time unit between 435 s and 680 s is shown in Fig. 5.19. In the figure, the amount of the produced RBE caused by the regenerative braking is similar in the two methods. However, when the driving strategy optimisation method is adopted, the produced RBE can only be efficiently utilized in the time unit 575–590 s. By using the integrated optimisation, RBE is high-efficiency reused in several time units because the traction/braking regimes are frequently matched during the train operation.

Next, the results of the proposed method is compared with the optimal solutions of the cooperative control problem calculated by the DP algorithm. As shown in Table 5.5, although the net energy consumption of the improved MARL increases by 0.6%, 1.0%, and 3.5% compared to the DP algorithm in three cases, the computing time noticeably diminishes by 17.3%, 60.1% and 97.9%, respectively.

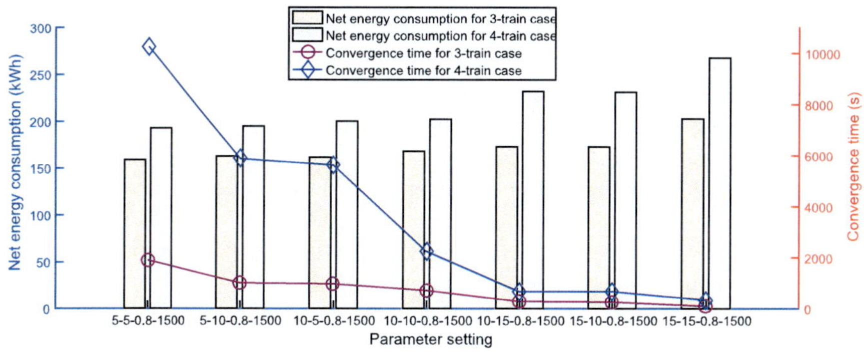

(a) Change the speed interval and the distance interval

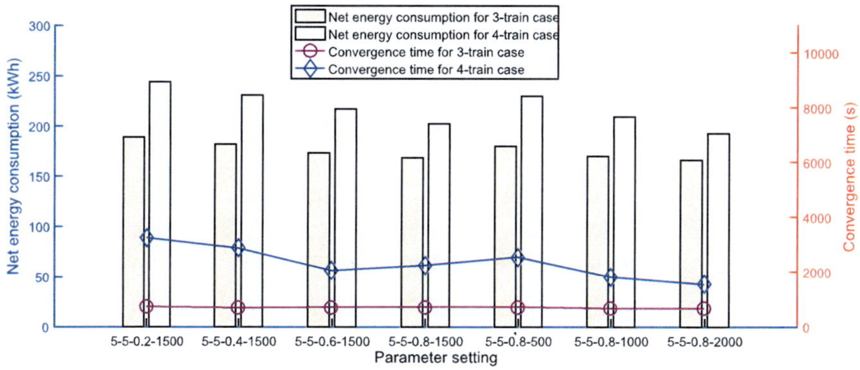

(b) Change the energy conversion rate and the maximum transition range of RBE

Fig. 5.20 Results of the sensitive analysis

Although the whole state space of all trains is searched and the optimal solution is obtained by using the DP algorithm, the calculation amount will exponentially grow when the number of operated trains increases, which implies a longer computing time. By applying the MARL algorithm, the local Q-functions are updated by iteration. In each episode, only the state transitions among part of the states are sampled to update the local Q-functions. Although it will take long time to obtain the optimal policy by using the MARL algorithm, a satisfying solution (i.e., approximate optimal solution) can be obtained within finite iterations in a shorter time compared to the DP algorithm.

Then, a sensitive analysis is done to illustrate the impacts of the parameter changes on the performance of the improved MARL algorithm. The results of the sensitive analysis is shown in Fig. 5.20. In the figure, the sequence 5–10-0.8–1500 means the speed interval is 5 km/h, the distance interval is 10 m, the energy conversion rate φ is 0.8, the maximum transition range $\Delta x^{\max}$ is 1500 m m.

As shown in Fig. 5.20(a), when the interval of speed or distance reduces, the net energy consumption will decrease and the convergence time will increase. The better solution quality is due to that there will be more optional driving strategies in the STS network with a smaller interval. Therefore, trains have the potential to choose better driving strategies in the cooperative control. Besides, the larger state space and the action space caused by the smaller interval lead to more calculation amount in each iteration.

In Fig. 5.20(b), the net energy consumption will decrease with more energy conversion rate. This is because more RBE can be produced by regenerative braking and reused by other trains. A similar tendency is also found when increasing the maximum transition range of RBE, for the reason that the RBE can be transmitted among trains at more joint states. Besides, the computing time only makes a small change in this set of sensitive analysis, because the spaces of state and action remain the same and only the reward is changed according to different parameters.

5.5 Conclusions

In this chapter, the train timetabling optimisation approaches considering the reuse of RBE were introduced for energy-efficient train operation. The principles of energy saving, the modelling of optimisation approaches, and the corresponding algorithms were discussed. In the first approach, a numerical algorithm was designed to adjust the departure time of trains for a better usage of RBE. This approach can efficiently reduce the practical energy consumption for metro systems and does not influence the punctuality of trains. Some examples based on the operation data of Beijing Yizhuang Metro Line showed that the cooperative model can save 2.4% of energy for one trip and 0.47% for one peak hour operation compared with the optimal train control approach, which illustrated the energy-saving effectiveness of the proposed model. This timetabling approach can be off-line applied to make the existing train timetable further has better energy-saving effect. In addition, the computation time is short enough to apply the algorithm to the real-time control system. In this way, the departure time can be adjusted according to the real-time states of the train.

The second approach is an integrated optimisation approach to jointly optimise the train timetable and driving strategy. The traction/braking regimes during the inter-station train operation were matched, to get better energy-saving effects. According to the examples based on the real-world data, the net energy consumption by using the integrated optimisation method was reduced by up to 16.5% compared to the driving strategy optimisation method.

The approaches introduced in this chapter focused on the metro system, where homogeneous trains with the same characteristics and stop patterns are operated over a double-track rail line with a fixed speed limit throughout. In mainline railway, there are more complex train operation scenarios, including heterogeneous trains, different stop patterns, single or multiple tracks, multiple platforms, etc. Therefore, it is necessary to extend the integrated optimisation approaches to these more complex scenarios in the future.

References

1. Busoniu L, Babuska R, De Schutter B (2008) A comprehensive survey of multiagent reinforcement learning. IEEE Trans Syst Man Cybernetics Part C (Appl Rev) 38(2):156–172
2. Hansen I, Pachl J (2014) Railway timetabling and operations: analysis, modelling, optimisation, simulation, performance, evaluation. Eurail Press
3. Kok JR, Vlassis N (2006) Collaborative multiagent reinforcement learning by payoff propagation. J Mach Learn Res 7:1789–1828
4. Luan X, Wang Y, De Schutter B, Meng L, Lodewijks G, Corman F (2018) Integration of real-time traffic management and train control for rail networks-part 2: extensions towards energy-efficient train operations. Transp Res Part B: Methodol 115:72–94
5. Mo P, D'Ariano A, Yang L, Veelenturf LP, Gao Z (2021) An exact method for the integrated optimization of subway lines operation strategies with asymmetric passenger demand and operating costs. Transp Res Part B: Methodol 149:283–321
6. Scheepmaker GM, Goverde RM, Kroon LG (2017) Review of energy-efficient train control and timetabling. Euro J Oper Res 257(2):355–376
7. Su S, Wang X, Cao Y, Yin J (2019) An energy-efficient train operation approach by integrating the metro timetabling and eco-driving. IEEE Trans Intell Transp Syst 21(10):4252–4268
8. Su S, Tang T, Roberts C (2014) A cooperative train control model for energy saving. IEEE Trans Intell Transp Syst 16(2):622–631
9. Su S, Li X, Tang T, Gao Z (2013) A subway train timetable optimization approach based on energy-efficient operation strategy. IEEE Trans Intell Transp Syst 14(2):883–893
10. Su S, Wang X, Tang T, Wang G, Cao Y (2021) Energy-efficient operation by cooperative control among trains: a multi-agent reinforcement learning approach. Control Eng Practice 116:104901
11. Tian Z, Hillmansen S, Roberts C, Weston P, Zhao N, Chen L, Chen M (2016) Energy evaluation of the power network of a DC railway system with regenerating trains. IET Electrical Syst Transp 6(2):41–49
12. Tian Z, Zhao N, Hillmansen S, Roberts C, Dowens T, Kerr C (2019) SmartDrive: traction energy optimization and applications in rail systems. IEEE Trans Intell Transp Syst 20(7):2764–2773
13. Watkins CJ, Dayan P (1992) Q-learning. Mach Learn 8(3):279–292
14. Zhou L, Tong LC, Chen J, Tang J, Zhou X (2017) Joint optimization of high-speed train timetables and speed profiles: a unified modeling approach using space-time-speed grid networks. Transp Res Part B: Methodol 97:157–181
15. Wang P, Goverde RMP (2019) Multi-train trajectory optimization for energy-efficient timetabling. Euro J Oper Res 272(2):621–635
16. Wang X, Tang T, Su S, Yin J, Gao Z, Lv N (2021) An integrated energy-efficient train operation approach based on the space-time-speed network methodology. Transp Res Part E: Logistics Transp Rev 150:102323
17. Wang P, Bešinović N, Goverde RMP, Corman F (2022) Improving the utilization of regenerative energy and shaving power peaks by railway timetable adjustment. IEEE Trans Intell Transp Syst 23(9):15742–15754
18. Yang X, Chen A, Ning B, Tang T (2017) Bi-objective programming approach for solving the metro timetable optimization problem with dwell time uncertainty. Transp Res Part E: Logistics Transp Rev 97:22–37
19. Yin J, Tang T, Yang L, Gao Z, Ran B (2016) Energy-efficient metro train rescheduling with uncertain time-variant passenger demands: an approximate dynamic programming approach. Transp Res Part B: Methodol 91:178–210
20. Zhu Q, Su S, Tang T, Liu W, Zhang Z, Tian Q (2022) An eco-driving algorithm for trains through distributing energy: a Q-learning approach. ISA Trans 122:24–37

Chapter 6
Energy-Efficient Train Driving Considering Energy Storage Systems

Gonzalo Sánchez-Contreras, Adrián Fernández-Rodríguez, Antonio Fernández-Cardador, and Asunción P. Cucala

6.1 Introduction

Electrified railways generally use electric drives consisting of induction motors for their movement. There are two modes of operation: as motors themselves, fed by the grid and consuming the energy needed to drive the train, and in generator mode when the electric brake is used to reduce speed. The latter is known as regenerative braking, where the kinetic energy of the train is transformed into electrical energy that is generally returned to the catenary after feeding the auxiliary services (Fig. 6.1).

Regenerated energy can be harnessed at the moment if there is another train accelerating in the same electrical sector as the first one [1]. If there is no such synchronization in the trains, the energy is attempted to be returned to the grid if it is possible. If it is not possible, when the grid is direct current (DC), the regenerated energy that is not consumed by other trains typically must be wasted in on-board resistors to avoid catenary overvoltage problems.

In the case of alternating current (AC) systems, which are commonly used in long-distance lines, power can be transferred back to the grid if there is no train in the vicinity of the braking train. The voltage on the catenary increases when a train regenerates. However, the voltage limit is not usually exceeded if there are not several trains in the same power sector regenerating at the same time [2]. This coincidence

G. Sánchez-Contreras · A. Fernández-Rodríguez (✉) · A. Fernández-Cardador · A. P. Cucala
Institute for Research in Technology, ICAI School of Engineering, Comillas Pontifical University, 23 Alberto Aguilera Street, 28015 Madrid, Spain
e-mail: adrian.fernandez@iit.comillas.edu

G. Sánchez-Contreras
e-mail: gonzalo.sanchez@iit.comillas.edu

A. Fernández-Cardador
e-mail: antonio.fernandez@iit.comillas.edu

A. P. Cucala
e-mail: Paloma.cucala@iit.comillas.edu

S. Su et al., *Energy-Efficient Train Operation*, Lecture Notes in Mobility,
https://doi.org/10.1007/978-3-031-34656-9_6

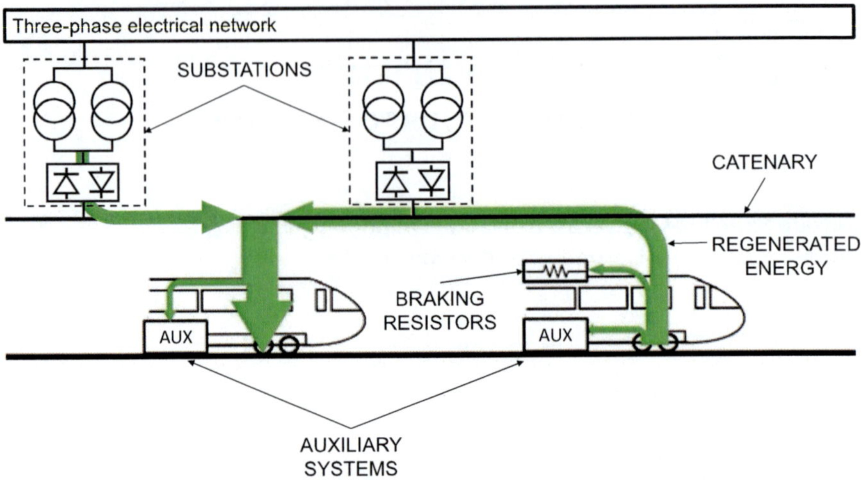

Fig. 6.1 Regenerated energy flow between trains

of trains is difficult to occur since the long-distance train frequency is not common to be as high as in metropolitan railways.

On the contrary, metropolitan lines are in most cases fed by DC systems. In these systems, the power cannot be returned to the grid through the substations because they have rectifier filters that do not allow the conversion from DC to AC. The use of regenerated energy is limited to cases where there is a train powering simultaneously with another train braking in the same electrical sector. If there are no other trains nearby and the train sends regenerated energy to the catenary, the voltage levels will rise over the allowed limits. To prevent this, the train has to dissipate, partially or totally, the braking energy in the brake resistors in the form of heat.

Thus, in DC grid railway systems, the amount of regenerated energy recovered depends on many factors: degree of train deceleration [3], type of driving [4], frequency of service (i.e. probability of train synchronization), power profile of the trains, traction network configuration, rolling stock, line voltage, terrain profile, length of electrical sectors, train auxiliary systems [5] and the size and configuration of the on-board accumulator if any [6].

The potential of braking energy in electrified railways is typically up to 40–45% of that consumed [7–9], and yet there are measurements showing only 19% recovery [10]. Thus, the loss of regenerated energy in the brake resistors on board trains could reach more than 25%. It can be minimised with the installation of reversible substations or energy accumulation equipment (at substations or onboard). Power inverters allow injecting the trains' energy regenerated by braking in the DC side back to the AC power grid [11] so that it can be used for other purposes. This strategy is currently used in different metro administrators of Spain such as Metro de Bilbao or Metro de Madrid. Power inverters are preferred in terms of energy efficiency. However, there is no economic profit of installing them if there is no payment for the energy returned to the net. On the contrary, accumulation systems do not have this limitation in addition to other benefits such as reduction of power peaks and their economic costs [12]. Furthermore, they can be used to use regenerated energy not only to feed traction demand but also for other purposes such as charging electric vehicles [13–16].

6.1.1 Accumulation Systems

Due to the problems presented by DC regeneration, it may be convenient, as already mentioned, to equip the system with devices that allow energy storage in the train itself or in the substations. Storage devices (ESS) can be batteries, flywheels, or capacitors. The advantage of these accumulators is that the regenerative energy does not have to be consumed immediately and it is not necessary the coincidence of a powering train with the braking of another in order to take advantage of the energy generated. This energy can be stored until another train demands it or the train itself needs it. As disadvantages, there are losses in the charging and discharging of the accumulator, and the energy storage capacity is limited. Other advantages provided by these devices are:

- Reduction of power peaks in the grid despite increased demand [17]. In the same way, energy storage can be used to increase the maximum usable power in DC systems [7]. This could be worthy not only from the technical point of view but also from the economical point of view reducing the power cost of the system under certain electricity billing frameworks.
- Voltage stabilization despite irregular demand profile [18]. It can be applied to stabilize the catenary voltage in weak networks to prevent excessive under voltage figures.
- Improvement of the energy efficiency of railway system. These systems increase the regenerative energy usage increasing the receptivity of the network and reducing the energy wasted in brake resistors on board the trains [19]. Thus, a reduction of the complete system energy consumption is achieved maintaining the same service quality. This energy reduction will depend on the location, sizing and technology of the ESS [20].
- Catenary free operation. ESSs installed on-board trains allow electric trains to run in non-electrified track sections. This catenary free operation is typical of trams where the preserving requirements for historical city centres could difficult the deployment of an electrical system in certain areas [21–23]. It is also typical of diesel-electric trains where the energy storage system allows the train to reduce its fuel consumption [24].
- Increase in substation distance. ESSs installed in fixed positions at the track can support the substations energy supply to trains allowing greater distances between substations [7].

Currently, there are two forms of storage: static accumulators usually located in substations, and on-board accumulators in the trains. While the energy flow of the former involves all moving trains in the electrical sector in which they are located, on-board accumulators only involve the internal energy flow of the train itself.

The economic investment of on-board accumulators is higher since several units are needed for each vehicle depending on the desired storage capacity. Moreover, the total train mass increases, requiring additional space in the trains [25]. However, this approach is the solution to be applied in catenary free operation scenarios and has other advantages compared to stationary accumulators, such as avoiding losses in power transmission over the catenary [7]. On the other hand, The development of energy accumulators has been widely studied; numerous studies try to find the best technology, configuration or size [26–28].

The selection of the appropriate type of accumulator must take into account the size, weight, efficiency and life cycle of the device [9]. The possibilities are: batteries, flywheels, double-layer capacitors (also known as supercapacitors), and superconducting magnetic energy storage (SMES). This last option is not frequently applied in railway systems nowadays.

Batteries have a limited number of charge and discharge cycles [25], as well as being heavy. Both characteristics make them less favorable for use in onboard trains compared with other technologies such as double-layer capacitor. However, references can be found in the literature where the application of batteries on board the train is analysed for energy saving [27] and catenary-free operation of trams [29]. On the contrary, the main advantage is the high energy density performance. That is why batteries are the most common technology for fixed ESSs at substations. Conventional lead-acid batteries have been used in railway application for a long time [30]. However, nowadays lithium-ion (Li-Ion) and nickel-metal (Ni-MH) batteries are the emerging technologies because they perform better in terms of energy density and life cycle [31].

Flywheels present high energy density and large power capacity but they have rather unfavorable dimensions for their use and high self-discharge ratios compared to batteries or supercapacitors. Besides there are some doubts about their safety in onboard applications. The placement of a flywheel on top of a moving vehicle in which people are traveling causes reluctance at the time of its implementation [9] although some onboard applications can be found such as the Rotterdam trams for catenary free operation [29]. Thus, flywheels are more suitable for accumulation in substations with more real examples for energy saving and voltage stabilization purposes [32].

Double-layer capacitors have a higher storage capacity than other traditional capacitors as well as lower internal resistance. They allow the formation of accumulators of acceptable dimensions and also adapt well to different voltages. They also stand out for their high efficiency, capacity for high dynamic load changes, and suitability for cyclic operation [33]. In addition, they have reduced maintenance cost, and their energy storage capacity can be scaled or cascaded. In this way, it is possible to adapt the energy storage individually to the specific needs of each train.

They also have a high power density, although lower energy density and higher self-discharge ratio compared to batteries [34]. Consequently, it is the system commonly used for energy storage on board light rail vehicles (LRV), also without catenary [35], achieving significant savings [36]. There are also studies about its benefits onboard diesel trains [37] or hybridized with batteries to combine the best characteristics of both technologies [38]. Despite the wide on-board applications of double-layer capacitors, there are also fixed applications at substations, for voltage stabilization and energy saving purposes [39], and commercial products such as the SITRAS SES (Stationary Energy Storage) developed by Siemens [40] that is installed in Spain (Madrid), Germany (Bochum, Cologne, Dresden and Koln) and China (Beijing) [41].

6.1.2 Efficient Driving and Regenerative Braking

As shown before, the use of regenerative braking is an energy efficiency measure with the potential for energy reduction. One may think that if it is possible to recover all the regenerated energy produced then it is not worth to optimise the train driving because the excess of energy consumed by a non-optimise speed profile will be returned to the grid by means of braking energy. However, this approach disregards that the excess of energy consumed by trains increases the losses in power transmission and due to electro-mechanical transformations in motor. Moreover, maximising regenerative energy usage needs the installation of accumulators and inverters, which have limitations in power or capacity. Therefore, energy-efficient driving will always be beneficial providing significant energy saving figures [42].

Several studies have analysed the train driving considering regenerative energy. In [43] the authors proposed to modify the braking curves to obtain braking periods with small changes in the schedule, increasing the regenerated energy by 22%, which they consider, would be fully assumable by the substations. On the contrary, in [44] the authors consider that the most efficient solution is to first optimise the speed profiles of each train and subsequently increase the responsiveness to regeneration. In [45] the aim is to optimise driving by selecting coasting points during the train journey. Regenerated energy is accounted for the optimisation process but later it is considered in the results analysis. The authors derive that energy saving capacity of efficient-driving is lower if there is regeneration between trains since the energy demand of the network is reduced. Furthermore, by adding coasting periods, there are fewer braking periods and from lower speeds, so less regenerated energy is produced. They conclude that the interaction between economic driving (in this case with coasting points) and regenerative braking is complicated to evaluate.

This limitation is solved using energy-efficient driving models that integrate regenerated energy. These models usually take as objective to minimise the net energy measured at pantograph, that is, subtracting the braking energy to the traction energy and auxiliary equipment consumption [46]. Regenerated energy can be reduced by a coefficient that represents the receptivity of the catenary to braking energy (that can be adjusted taking into consideration the synchronization probability of trains braking and powering, and the presence of devices to harness regenerate energy, such as inverters or accumulators). More complex studies aim to minimise train energy consumption and maximise the train synchronization [47–49]. However, in these cases, the driving is affected not only by considering regenerated energy but also by adapting its running time to achieve better synchronization rates.

Previous models do not contemplate on-board accumulation systems and the presence of stationary energy accumulators can be modelled modifying the receptivity of the net. However, other studies proposed models to obtain energy-efficient driving including the presence of on-board storage systems. For instance, in [35] the energy-efficient driving model for journey with catenary free sections is obtained while in [36] the model considers an on-board energy storage system for energy efficiency purposes. In [50], the optimal ATO speed profiles are obtained in different scenarios considering the presence of on-board energy storage and optimising the energy consumption estimated at substation level. The energy calculation includes transmission losses and different network receptivity depending on the traffic and the presence (or not) of inverters connected at substations. The authors conclude that the optimal driving for certain running times differs when there are big differences in braking energy usage. However, in high receptivity scenarios or when the train includes on-board storage capacity, the optimal ATO driving for a running time does not experiment modifications.

The purpose of this chapter is the introduction of the modelling of energy storage system for railways. In particular, models for on-board and track-side storage systems are addressed. Furthermore, the particularities of energy-efficient driving of metro ATO trains are detailed and a case study is presented to analyse the influence of storage system in the metropolitan trains efficient-driving.

The rest of the chapter is organized as follows: Sect. 6.2 presents the modelling of energy storage systems for railways, Sect. 6.3 introduces the efficient-driving in metro ATO trains, in Sect. 6.4 the results of a case study are analysed and the main conclusions are detailed in Sect. 6.5.

6.2 Modelling of Energy Storage Systems for Railways

This section is devoted to the modelling of energy storage systems in railways. First, on-board systems modelling is detailed and, then, it is addressed the modelling of track-side storage.

6.2.1 On-Board Energy Storage Systems

By installing an on-board energy accumulator, trains could accumulate their own regenerated energy while braking and use it during the next start, which would modify the energy flow in the train.

There are different control strategies that modify the behavior of the accumulator and the result in energy consumption. In particular, it can be differentiated 4 general control strategies [31]:

- Control strategy to maintain the maximum state of charge: It is suited for catenary free operation to ensure that the train arrives at the catenary gaps with the maximum energy available in the storage system.
- Control strategy to maintain the minimum state of charge: It is suited for energy saving purposes. Maintaining the minimum state of charge allows the on-board storage system to be available to receive as much regenerated energy reducing the probability that it has to be burnt in on-board resistors.
- Speed-based dependent control strategy: depending on the speed of the train, the on-board storage system could use one of the first two strategies. Thus, the train can arrive with the maximum energy stored at catenary-free sections maximising the energy savings in the rest of the line. It is suitable when there are different operational speeds in catenary and catenary-free sections.
- Look ahead strategy: It uses the route details to determine when to charge and discharge the on-board storage system combining the first two control strategies to minimise the energy consumption from the catenary while ensuring that the storage system has enough state of charge for catenary-free sections.

The model presented in this section is devoted to energy efficiency and fully electrified railway systems. Therefore, the control strategy to maintain the minimum state of charge is the basis for the model.

The authors of [36] suggest that the optimisation of running and the charging and discharging order of the accumulator should be done jointly. However, after performing the optimisation they conclude that the speed profile hardly changes with or without energy accumulator being the order for charging the accumulator when the train brakes and discharging when powering. Therefore, in efficient-driving studies the charging and discharging of the accumulator only depends on the power demand profile and its corresponding speed profile.

The energy consumption model for the efficient-driving to calculate the energy consumed at the pantograph level is described in Fig. 6.2 (where the maximum power of the accumulator is never exceeded).

When the train is braking the accumulator is charged with the regenerated energy not used by the auxiliary systems, if it is not already full (see the charging energy storage box in Fig. 6.2). If the storage is full or the regenerated power is over the maximum power of the storage system, the braking energy (or the part that cannot be accepted by the storage system) is sent to catenary or on-board resistors depending on the network receptivity.

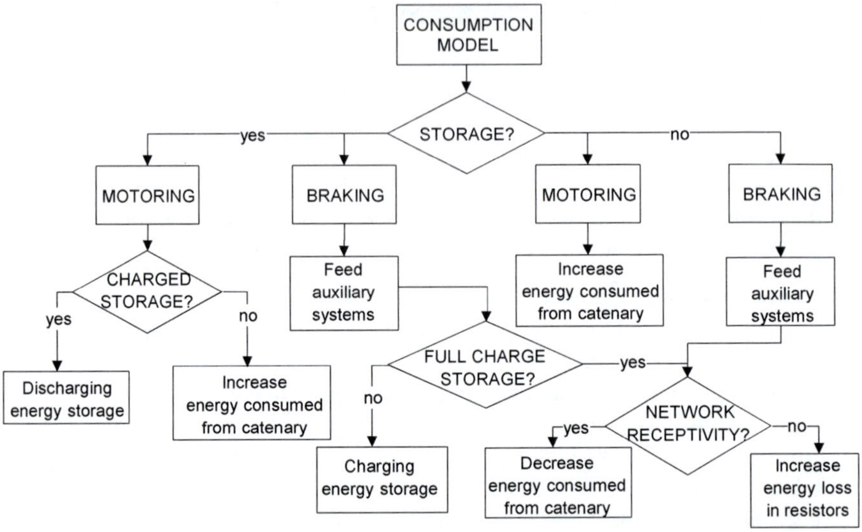

Fig. 6.2 Flow-chart of energy consumption model

On the contrary, when the train is accelerating it is first fed by the on-board accumulator if the state of charge is over the minimum (see the discharging energy storage box in Fig. 6.2) [51]. The train is also powered from the catenary if the energy supplied by the accumulator is not sufficient.

Figure 6.3 shows a simulation example of the operating cycle of the on-board storage system on the train during the travel of several inter-stations. During braking the kinetic energy of the train is stored in the storage system subtracting all the losses due to the transformation into electric energy and to the storage process (and the potential energy due to track grade variations). In the following start-up period and during traction, the train takes the required power from the storage system first and, in case it is not sufficient, from the network. This means that there is at least one charging and discharging cycle at each inter-station. The greater the train speed at the start of braking, the greater the regenerated energy, as can be deduced from the kinetic energy Eq. (6.1). Therefore, the charging and discharging control of the accumulator affects the energy consumed by the train.

$$E_k = \frac{1}{2} \cdot M \cdot \mathrm{v}^2 \tag{6.1}$$

where E_k is the kinetic energy, M is the train mass and v is the train speed.

The storage system considered is based on the most common technology for on-board applications and consists of supercapacitors. The model of the train circuit with an on-board accumulator is shown in Fig. 6.4.

The maximum energy accumulated by each device is W_M when the voltage is maximum (U_M) as shown in Eq. (6.2). However, since the minimum voltage is

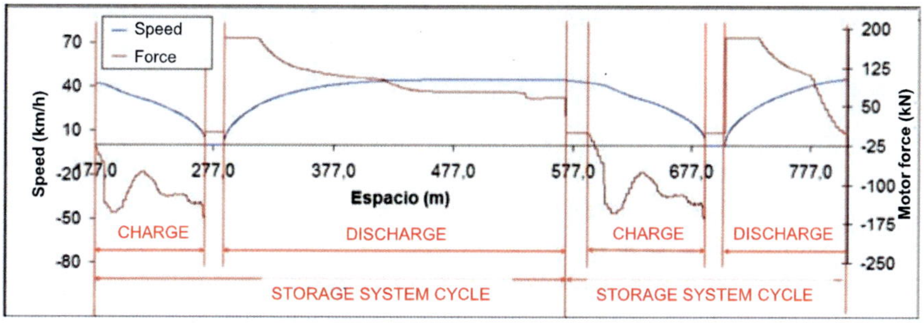

Fig. 6.3 Speed and motor force profile between stations and storage system cycle

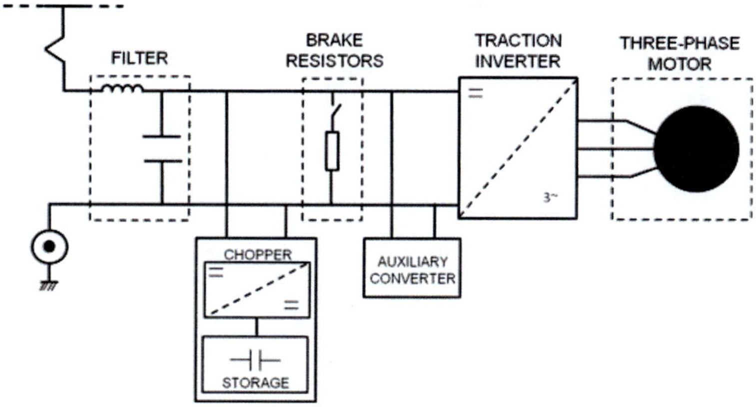

Fig. 6.4 Train circuit with on-board energy storage system

limited to U_m, the total energy accumulated that can be used, W_U, is calculated using Eq. (6.3). The efficiency of supercapacitors (η_c) is affected by the energy dissipated in the supercapacitor resistance (W_R), that depends on the discharge current (I), the ohmic resistance (R) and voltage limit represented by U_M and the discharge ratio (d), as shown in Eq. (6.4). Voltage discharge ratio can be calculated using Eq. 6.5 [52].

$$W_M = \frac{1}{2} \cdot C \cdot U_M^2 \tag{6.2}$$

$$W_U = \frac{1}{2} \cdot C \cdot \left(U_M^2 - U_M^2\right) \tag{6.3}$$

$$\eta_c = 1 - \frac{W_R}{W_U} = 1 - \frac{R \cdot C \cdot U_M \cdot I \cdot (100 - d)/100}{W_U} \tag{6.4}$$

$$d = \frac{U_m}{U_M} \cdot 100 \tag{6.5}$$

6.2.2 Track-Side Energy Storage Systems

A detailed analysis of the impact on energy consumption of installing a track-side energy storage system can be performed using a detailed simulation model, such as the one presented in Chap. 7, that incorporates a multi-train model and a load-flow model to represent the electrical network. Newton–Raphson algorithm is the most widely used approach to solve the load-flow model [53].

The network simulation, apart from incorporating substations, catenary and feeders modelling, must consider the energy storage system model. This model includes the efficiency ratio, the charging/discharging control and the storage capacity. The efficiency can be calculated using Eq. (6.4) as explained in the previous section, assuming a constant efficiency value [54] or with a detailed model that states the internal resistance as a function of the state of charge [55].

The charging/discharging control usually depends on the voltage of the energy storage connection point. Generally, it follows the shape of the curve depicted in Fig. 6.5, where six parameters can be differentiated:

- $Ic_{\max}$ is the maximum charging current.
- $Id_{\max}$ is the maximum discharging current.
- $Vc1$ is the voltage value above the charging process starts
- $Vc2$ is the voltage value above the charging process achieves the maximum current

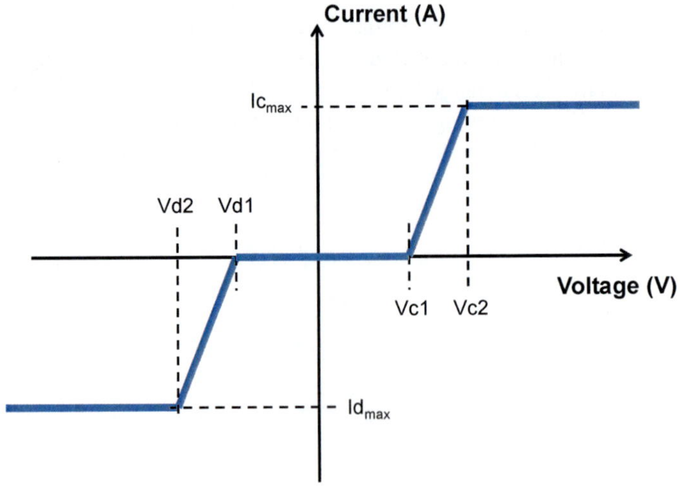

Fig. 6.5 Control curve for track-side energy storage system

- $Vd1$ is the voltage value below the discharging process starts
- $Vd2$ is the voltage value below the discharging process achieves the maximum current

The six parameters of the control curve are not constant and they are not linked one each other. This way, the performance of the power electronics that control the energy storage system can implement any strategy using these parameters depending on the catenary voltage level such as energy saving, voltage control, etc. The previous control curve can be defined by a set of piecewise functions. However, to avoid convergence problems in the load-flow algorithm, it can be approximated by a sum of sigmoid functions to boost the calculation time [56].

Apart from that control curve, other control conditions must be implemented to avoid the energy storage system absorbing more energy than its maximum level and to avoid that it provides energy when the minimum state of charge is reached.

6.2.2.1 Network Performance Coefficients

The previous on-board storage system model and energy consumption model (described in Sect. 6.2.1) can be incorporated into typical train driving simulation models for efficient-driving purposes. However, including track-side storage systems requires detailed electrical network models and multi-train simulation, as presented in Chap. 7. Such detailed models increase the complexity and calculation burden needed to produce energy-efficient train driving. A simplified network model based on coefficients can be included in the train model to include the network effect in the calculation of optimal driving based on the average performance.

The traction network can be characterised by means of two coefficients: the energy recovery coefficient and the loss coefficient. Both can be derived from the energy consumed in substations of all trains running on the line and depends on the traffic and the presence of devices such as track-side accumulators or inverters.

The energy produced by regenerative braking and not used by the train's auxiliary systems (E_{aux}), or the possible on-board accumulator (E_{sto}), is the regenerated energy available (E_{reg}). Although auxiliary systems consumption is variable, it is commonly modelled as a constant average value. Depending on the receptivity of the grid, part of this energy may go to catenary (E_{reg_cat}), while the rest will be burned in resistor banks on-board the train (E_{loss_ohm}). The E_{reg_cat} energy can be used by other trains, or sent to reversible substations or accumulated by track-side storage systems.

The amount of energy that substations must supply can be reduced by regenerated energy. In any case, the regenerated energy initially produced is greater than that available to be used by another train or accumulation device since different losses and efficiencies arise in the motors, power converters and energy transport.

Therefore, a coefficient is needed to characterise the total use of regenerated energy in the grid so that it is possible to compare different scenarios. The modelled coefficient is the energy recovery coefficient (ER) that can be calculated using Eq. (6.6).

The ER quantifies the energy savings measured at substations in a given scenario divided by the energy regenerated by all trains and available to be injected into the grid.

$$ER = \frac{E^b_{cons\_sub} - \left(E_{cons\_sub} - E_{reg\_sub}\right)}{\sum_t E^t_{reg}} \tag{6.6}$$

where $E^b_{cons\_sub}$ is the substation energy consumed by all trains if no regenerated energy can be sent to catenary, E_{cons_sub} is the net energy consumed at substation by all trains in the scenario analysed, E_{reg_sub} is the energy regenerated by all trains that is used by inverters or track-side storage systems and $\sum_t E^t_{reg}$ is the total energy regenerated by all trains t measured at pantograph.

The net energy consumed in substations is calculated as the energy provided by the substations (E_{cons_sub}) in the scenario considered, minus the energy that could have been fed into the AC grid through power inverters or stored in track-side storage systems if the substation had them (E_{reg_sub}). Savings are calculated with respect to a base case with a non-receptive network ($E^b_{cons\_sub}$). If a train t is braking in the base case, the regenerated energy can only be used to power the train's auxiliary systems (and the on-board energy storage if available). The rest of the energy would have to be burned in on-board resistors. If, on the other hand, the train were not braking, both the traction energy and that of the auxiliary systems would be provided by the power grid or by the train's on-board accumulator.

The use of regenerated energy, defined by ER coefficient, depends on different factors such as traffic and the infrastructure. Thus, with dense traffic, the probability of coincidence of trains powering and braking is higher and, as a consequence, the value of ER will be greater. On the other hand, the infrastructure also affects the value of ER because the more connected track sections, the more capacity to transfer regenerated energy between trains. Moreover, the presence of devices like track-side energy storage systems or inverters in substations increases the usage of regenerated energy and, therefore, the value of ER.

The loss coefficient (p) is defined in Eq. (6.7). This coefficient characterises the energy lost in the catenary and in the substations themselves, considering the total energy demand in pantograph ($\sum_t E^t_{cons}$) with respect to the energy provided by the substations in the base case ($E^b_{cons\_sub}$).

$$p = 1 - \frac{E^b_{cons\_sub}}{\sum_t E^t_{cons}} \tag{6.7}$$

From the traction energy demanded by a train, and the available energy it has regenerated (both measured in pantograph), it is possible to calculate an estimation of the consumption of that train in substations (ES^t_{cons}). It is sufficient to contemplate the electrical losses and the recovery of regenerated energy by means of the coefficients defined above as shown in Eq. (6.8). The first term of the equation calculates the energy consumed in substations, assuming zero receptivity of regenerated energy.

The second term decreases this consumption with the regenerated energy that has actually been used in the substations, reducing the net consumption.

$$ES^t_{cons} = (1 + p) \cdot E^t_{cons} - ER \cdot E^t_{reg} \tag{6.8}$$

6.3 Energy-Efficient Driving in Metro ATO Trains

Modern metropolitan lines apply Automatic Train Operation systems (ATO) to drive the trains automatically. The departure from stations starts with the reception of a set of driving commands from the centralized regulation system that are processed by the on-board ATO equipment. Typically, the driving command sets are pre-programmed in the ATO track-side equipment at each station and consist of four parameters: holding speed (vh), coasting speed (vc), re-motoring speed (vr), and braking rate (b) [57]. These parameters produce two kinds of driving styles, as shown in Fig. 6.6: holding speed driving and coasting-re-motoring driving.

Coasting-re-motoring driving is performed when vh is equal to 0, and vc and vr are different from 0. In this kind of speed profile, the train applies maximum traction effort from the departure until it reaches the coasting speed. At this moment, the traction is set to 0 to start coasting. From this point, the train coasts until its speed fall by more than the re-motoring speed from vc. Then, maximum traction effort is applied again until the train's speed reaches vc again. Maximum traction and coasting commands can be interrupted to use the braking effort needed to fulfill speed limitations and braking curves up to the station. For instance, at the end of the second coasting period in Fig. 6.6b, the train applies braking effort to face the maximum speed reduction.

On the other hand, holding speed driving is performed when vh is different from 0, and vc and vr are equal to 0. This driving style forces the train to reach and maintain a speed equal to vh. As in the previous case, braking effort is applied to fulfill speed limitations and braking curves up to the station.

In both cases, driving command b defines the deceleration that the train will apply in the final braking phase.

After the departure, the ATO driving logic drives the train to the following station performing a specific speed profile resulting from the driving commands received and the track characteristics as input data. Complex algorithms are part of the ATO driving logic. They are in charge of different functions: braking curves calculation (with variable deceleration depending on the track grade variations to maintain the train braking level), speed regulation (according to maximum speed profile, holding speed, and braking curves), start braking points calculation (to ensure that the train can follow the braking curve and make a smooth transition from traction to the braking curves), start coasting points calculation or passenger comfort control (limiting the jerk and imposing a hysteresis cycle between traction and braking mode) [58, 59].

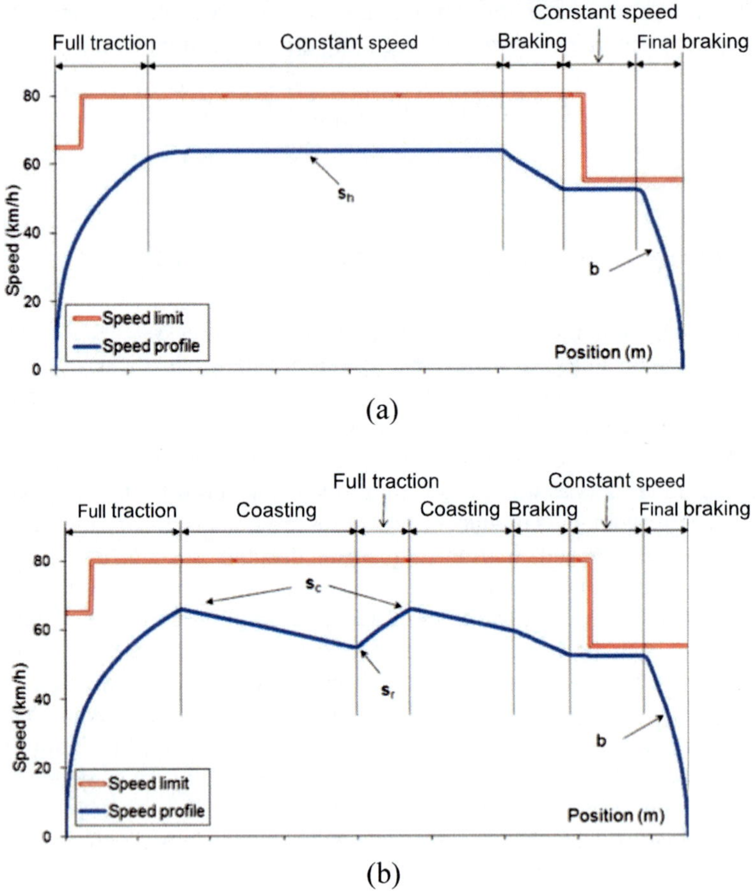

Fig. 6.6 ATO driving commands in two speed profiles: holding speed driving (**a**) and coasting-re-motoring driving (**b**)

Thus, the efficient-driving problem in ATO systems is the design of the driving commands to be programmed. Several sets of driving commands are programmed at each station, typically four: the fastest one (to recover delays), the one to perform nominal running time (to fulfill the timetable) and two slower sets of driving commands (to increase the distance with the preceding train if it is delayed). The driving commands sets must be designed off-line, tested in the train and then programmed in the track-side equipment.

In this regard, although many other techniques can be applied, multi-objective nature inspired algorithms seem well suited to this problem. They are flexible enough to include the complex and non-linear ATO driving logic. Besides, there are two conflicting objectives (running time and energy consumption) and the result is a Pareto front containing the most efficient speed profiles for the complete range of possible running time values. In view of the Pareto front, the designer can select the most appropriate sets of driving commands to be programmed taking into account trade-offs between energy and running time.

Among the different nature-inspired techniques that have been applied, Multi-Objective Particle Swarm Optimisation (MOPSO) has been demonstrated as an efficient method to obtain the Pareto front of optimal speed profiles [60].

This algorithm applied to the efficient-driving problem in ATO trains aims to minimise at the same time the running time and the contribution of the train to the energy consumed at the substations. Thus, the following objective function can be defined:

$$\min f(\hat{x}) = \min\left[RT^t(\hat{x}), ES^t_{cons}(\hat{x})\right] \tag{6.9}$$

where $RT^t(\hat{x})$ is the running time performed by the train using the set of driving commands $\hat{x}$ and $ES^t_{cons}(\hat{x})$ is the fraction of the energy consumed at the substations (calculated using Eq. (6.8) shown in Sect. 6.2.2.1) due to the train driving using the set of driving commands $\hat{x}$.

This algorithm uses a population of individuals to search for solutions. Each individual of the population is a particle (i) and the population of particles is a swarm. Particles fly modifying their position inside the search space seeking the best results. A particle's speed ($\hat{v}_i$) and position ($\hat{x}_i$) is modified iteratively using its best individual position found ($\hat{p}_i$) and the best position of the complete swarm ($\hat{p}_g$).

Every particle of the swarm represents a vector of decision variables $\hat{x}_i = (vc, vr, vh, b)$ containing the set of driving parameters for the ATO commands. The position of the particles of the swarm is updated at each iteration (n) according to Eqs. (6.10) and (6.11). The influence of the personal and the swarm experience is modulated through the social factors c_1 and c_2, respectively, and r_1 and r_2 are random numbers between 0 and 1. The coefficient w represents the inertia weight of the particle and decreases from an initial w_1 to w_2, helping the swarm search based on global experiences on the first iterations to search in new areas and progressively orient themselves towards their local experience to discover optimal points in their vicinity.

$$\hat{v}_i(n) = w\hat{v}_i(n-1) + c_1 r_1\left(\hat{p}_i - \hat{x}_i(n-1)\right) + c_2 r_2\left(\hat{p}_g - \hat{x}_i(n-1)\right) \tag{6.10}$$

$$\hat{x}_i(n) = \hat{x}_i(n-1) + \hat{v}_i(n) \tag{6.11}$$

Once the position of the particles is updated, the running time (RT^t) and energy consumption (ES^t_{cons}) of all of them is calculated. Those values are calculated making use of a train simulation model. This model could take into account the use of on-board energy storage systems, including the model described in Sect. 6.2.1, or the presence of track-side energy storage systems applying the network coefficients presented in Sect. 6.2.2.1.

After that, the non-domination criterion is applied to classify the fitness of the particles. A particle is considered non-dominated if there is no other particle with, at the same time, better energy and running time values. Furthermore, a crowding distance mechanism [61] is applied to select the $\hat{p}_g$ from the non-dominated positions of the swarm.

Finally, the best sets of driving commands found are updated incorporating the non-dominated solutions identified in the iteration and discarding the previous solutions that the new ones dominate.

6.4 Case Study

The models presented have been applied to a real railway system from Metro de Madrid to analyse the influence of storage systems in train efficient-driving. This metro line comprises 18 stations across 13.5 km, with gradient profiles up to 50 mm/m. The total trip time of a cycle is close to an hour. The electrical network consists of a rigid overhead with 1500 DC supply and six interconnected substations, which allows for better energy transfers between trains in the line.

The traffic regulation system of the considered metro line establishes that the speed profiles that can be programmed to dictate the driving of the trains is a set of four profiles. This set starts with the flat-out speed profile or number 0, which is considered the fastest one, increasing the count with slower speed profiles up to number 3. Accordingly, the difference between the maximum running time of profiles 0 and 3 is limited to 20 s. The headway between two consecutive trains is 2 min for peak hours and 15 min for off-peak hours.

The on-board energy storage device modelled for the case study is based off the "MITRAC Energy Saver" from Bombardier [25]. This device has been in use in a LRV on Mannheim since September 2003, and it also has been used for simulations in a European metropolitan system [25]. The considered energy storage device has a total capacitance of $C = 37.5\,\mathrm{F}$ and a maximum stored energy of $W_M = 1.296\,\mathrm{kWh}$ per device. The total available amount of energy (the difference of state of charge with maximum voltage U_M and minimum voltage U_m) is $W_U = 0.972\,\mathrm{kWh}$, with maximum and minimum voltages of $U_M = 500\,\mathrm{V}$ and $U_m = 250\,\mathrm{V}$.

The train is modelled with two trailers and four motor cars of 321 kW each. Each motor car has two powered bogies with two motors per bogie, up to a total of 16 motors of 80.25 kW. The storage device has the capability of delivering 400 kW of maximum power, considering its maximum current output of 800 A [62], which covers the energy demand of one motor car and the auxiliary systems. To fully power the train, four storage devices are considered, which increases the total mass by $M = 4 \times 477 = 1908\,\mathrm{kg}$.

Regarding efficiency, the ultracapacitors have been modelled with a varying value between 75 and 97%, whereas the bidirectional converter has been assumed constant and equal to 95% [36].

Table 6.1 Initial and final charge analysis for one interstation

Initial charge (%)	Final charge (%)	Running time (s)	Traction energy (kWh)	Braking energy (kWh)	Energy consumption (kWh)	Stored energy (kWh)	Wasted energy (kWh)
0.00	48.19	95.70	12.93	3.72	12.27	3.39	– 0.01
25.00	47.38	95.70	12.93	3.72	10.95	3.69	0.00
35.00	47.62	95.70	12.93	3.72	10.44	3.69	0.00
45.00	47.24	95.70	12.93	3.72	9.89	3.69	0.00
55.00	47.65	95.70	12.93	3.72	9.39	3.69	0.00
65.00	47.24	95.70	12.93	3.72	8.84	3.69	0.00
75.00	47.27	95.70	12.93	3.72	8.32	3.69	0.00
85.00	47.56	95.70	12.93	3.72	7.80	3.69	0.00
95.00	47.22	95.70	12.93	3.72	7.25	3.69	0.00

6.4.1 Initial Charge Estimation

To estimate the initial charge for the on-board energy storage device at the start of each interstation, two simulations with different initial charges have been carried out. These simulations have obtained an estimation of the charge of the energy storage device, which is needed to design the set of ATO speed profiles in scenarios with an on-board energy storage device installed. Table 6.1 shows the value of the on-board energy storage device charge for one interstation with different initial charge conditions. These results reveal that similar amount of energy is stored in the device at the end of each interstation trip, proving the initial charge is completely consumed at the start and the final state of charge is nondependent of the initial value.

On the other hand, Table 6.2 shows the initial charge values for all the inter-stations of the line using a flat-out speed profile starting the trip with the storage system fully charged. These values will be used as input data in the efficient-driving design.

6.4.2 Scenarios Analysed

The case study has been analysed with 12 different scenarios that have been simulated with a fixed load of 30% of maximum passengers' load. The considered scenarios are presented in Table 6.3. The scenarios have been divided into two groups: off-peak hours and peak hours, with different recovery coefficients and a constant value for the energy losses coefficient.

Scenarios 1 and 7 do not leverage the energy regenerated through braking after feeding the auxiliary systems, whereas the remaining scenarios consider regenerated energy transfer flows.

Table 6.2 Storage charge before train departure with flat-out

Departure station	Initial charge (%)	Departure station	Initial charge (%)
VA1	100.00	M2	62.82
SC1	88.50	AR2	56.06
VC1	67.54	VR2	89.30
CI1	70.57	PE2	90.62
SF1	92.01	C2	55.60
DO1	63.86	S2	90.01
AL1	64.59	LV2	88.89
L1	47.24	EM2	88.75
DL1	62.46	PF2	76.57
PF1	45.24	DL2	88.92
E1	51.28	L2	87.77
LV1	43.57	AL2	47.08
S1	61.14	DO2	62.21
C1	59.37	SF2	89.95
PE1	90.32	CL2	81.10
VR1	61.87	VC2	67.29
AR1	62.82	SC2	71.34

Table 6.3 Considered scenarios

	Scenario	Reg. energy used by other trains	On-board storage device	Track-side storage devices	Recovery coefficient (%)	Energy losses coefficient (%)
Off-peak hours	1				0	2
	2	X			70	2
	3	X		X	90	2
	4		X		0	2
	5	X	X		70	2
	6	X	X	X	90	2
Peak hours	7				0	2
	8	X			95	2
	9	X		X	100	2
	10		X		0	2
	11	X	X		95	2
	12	X	X	X	100	2

In scenarios 2 and 8, the regenerated energy can be transferred between trains, but no storage devices are used. Scenarios 3 and 9 are similar but with track-side storage devices that increase the recovery coefficient. Trains are equipped with on-board storage devices in the remaining scenarios. In scenarios 4 and 10, the trains cannot return the regenerated energy to the catenary and, therefore, will obtain the same results. However, scenarios 5 and 11 consider energy transfer between trains, and scenarios 6 and 12 include the presence of track-side storage systems that increase the recovery coefficient.

6.4.3 Efficient-Driving Design

The efficient-driving design is performed applying the MOPSO algorithm presented in Sect. 6.3 to each interstation of the line and each scenario. Thus, the result obtained is a Pareto curve with the most efficient speed profiles for each case. Furthermore, it is considered a CBTC driving equipment where the driving commands can take continuous values between these ranges:

- Holding speed (vh) from 80 to 30 km/h
- Coasting speed (vc) from 80 to 30 km/h
- Re-motoring speed (vr) from 80 to 30 km/h
- Braking rate (b) from 0.8 to 0.6 m/s^2

A comparison of the Pareto optimal curves obtained in the different scenarios 1 and 2, which belong to off-peak hours, and 8, which corresponds to peak hours, is shown in Fig. 6.7. These curves present the different optimal speed profiles and their running times and energy consumption when there is no energy transfer between trains and when regenerative energy is considered. Allowing other trains to utilize part of the regenerated energy reduces the total energy consumption for the trip compared to scenario 1. Consequently, the designed speed profiles from scenario 1 are dragged down the Y-axis proportionally to the recovery coefficient considered.

Examining the Pareto fronts for scenarios 2 and 8, it can be observed that the optimal design of speed profiles for both scenarios is the same, with the only difference between both scenarios being the recovery coefficient applied to the regenerated energy. Therefore, the ATO commands designed for one time period still being optimal for other different periods, given the rest of the parameters are equal.

As can be seen, all the Pareto curves present a gap in the running time interval between 142 and 144 s. The absence of optimal speed profiles in this interval can be explained by the non-linear rules incorporated in ATO driving logic that can produce abrupt changes in the driving pattern with a small change in the driving command [57]. A change in the driving pattern makes all the solutions with a running time value in the mentioned interval present an energy consumption greater than the solution with a running time equal to 142 s, which dominates them, producing the gap. From 144 s the solutions present energy consumption values lower than the solution with

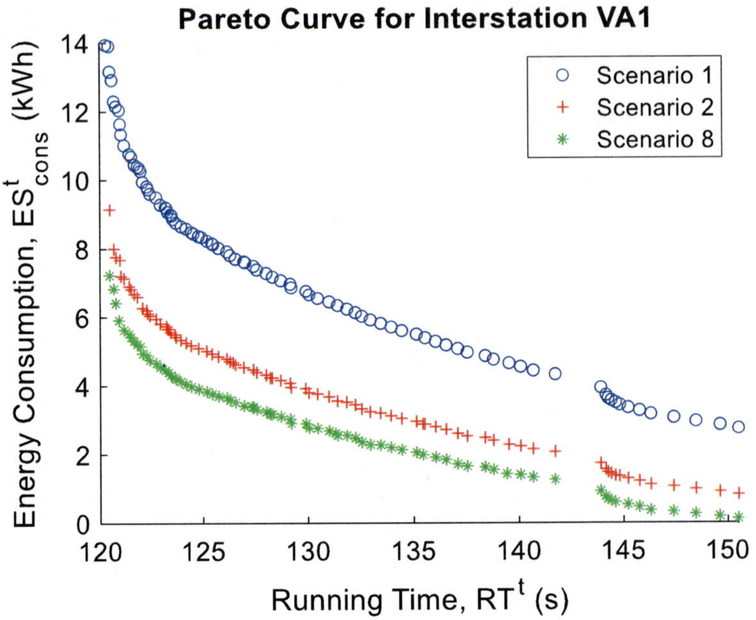

Fig. 6.7 Scenarios 1, 2 (off-peak hours) and 8 (peak hours)

a running time equal to 142 s and, therefore, are non-dominated and included in the Pareto front.

Nonetheless, when different receptivity scenarios for regenerative energy are compared, this statement is not always true. An example of this is shown in Fig. 6.8, where the Pareto fronts for scenarios 7, 8 and 9 are represented, as well as the Pareto curve for scenario 7 simulated with the parameters of scenarios 8 and 9. When the optimal set of speed profiles obtained for scenario 7 are evaluated in scenarios 8 (scenario 7_8) and 9 (scenario 7_9), several of the speed profiles which originally were located in the Pareto front no longer belong to the optimal curves for the other scenarios. Therefore, a new design of speed profiles is needed to ensure the optimality of the whole Pareto set.

The scenarios that have an on-board energy storage device installed in the trains (4, 5, 6, 10, 11 and 12) obtained the same optimal set of speed profiles than those scenarios without receptivity (1 and 7). The reason behind this result lies in the utilization of the regenerated energy from braking to charge the on-board devices, where in these scenarios the designed speed profiles barely feed any energy back to the network. These results show that scenarios 1, 4, 5, 6, 7, 10, 11 and 12 have an optimal set of speed profiles (where regenerative energy is mostly stored in the on-board accumulators) that is not completely equal to the optimal set of scenarios 2, 3, 8 and 9 (where the energy is injected into the network).

Therefore, minimising the traction energy consumption [58] (scenarios 1 and 7) to design the optimal set of speed profiles obtains driving commands that are valid

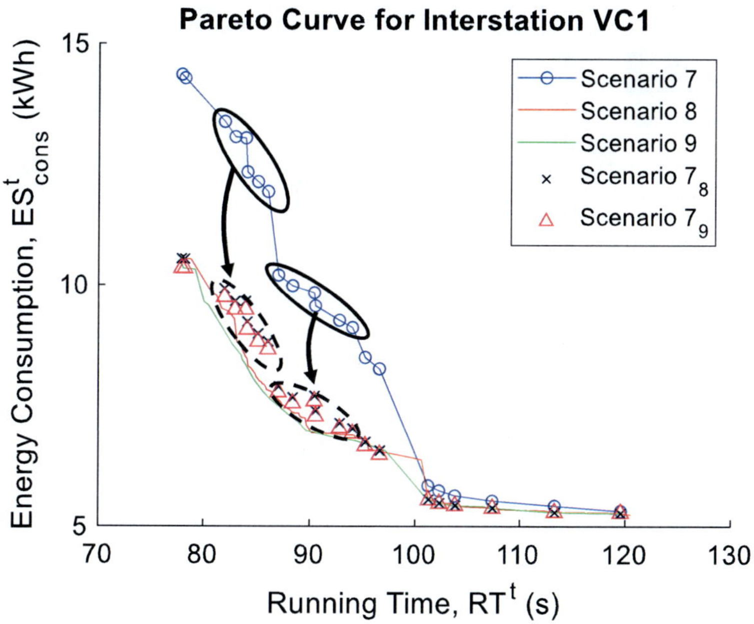

Fig. 6.8 Pareto curves for VC1 in scenarios 7, 8 and 9; and Pareto speed profiles for scenario 7 evaluated in scenarios 8 (scenario 7_8) and 9 (scenario 7_9)

for other scenarios with on-board energy storage devices. Several studies reached a similar conclusion [50]. Furthermore, due to the complete use of the accumulator charge during the interstation trips, the final state of charge is independent of the initial charge for the line.

6.4.4 Achievable Energy Savings Due to Efficient-Driving

The total energy consumption (ES^t_{cons}) for a complete cycle in off-peak hours and peak hours is shown in Figs. 6.9 and 6.10, respectively. These figures show a comparison between the energy consumed during a trip with a flat-out speed profile versus the nominal speed profile. The nominal speed profile at each interstation is taken from the Pareto curves as the speed profile with the minimum energy consumption whose running time is at most 5% greater than the minimum running time.

As can be seen, the application of efficient-driving provides energy reductions in all the scenarios studied. The energy saving result can be calculated in each scenario as the difference in energy consumption between the flat-out and the nominal driving divided by the energy consumed by the flat-out. The total energy savings obtained by each scenario can be seen in Table 6.4.

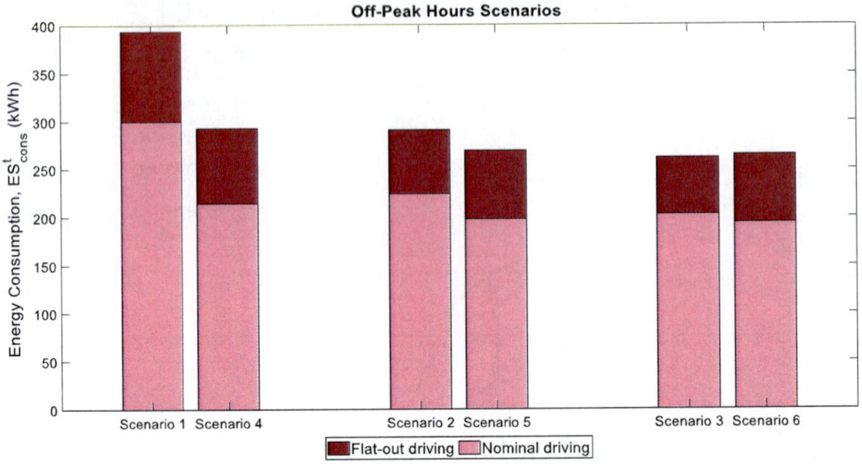

Fig. 6.9 Energy consumption in substations in off-peak hours for the flat-out and nominal driving

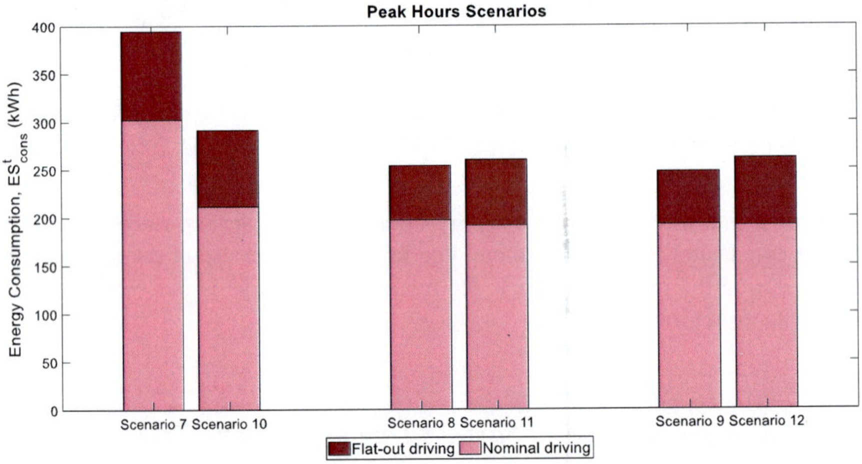

Fig. 6.10 Energy consumption in substations in peak hours for the flat-out and nominal driving

Table 6.4 Energy savings due to efficient-driving

Off-peak hours	(%)	Peak hours	(%)
Scenario 1	23.79	Scenario 7	23.36
Scenario 2	22.97	Scenario 8	22.28
Scenario 3	22.57	Scenario 9	22.24
Scenario 4	26.77	Scenario 10	27.36
Scenario 5	26.47	Scenario 11	26.43
Scenario 6	26.67	Scenario 12	26.89

Scenarios which do not take advantage of regenerative energy to minimise their consumption (1 and 7) present savings up to 23.36%, accomplished by driving with a designed nominal speed profile (number 1) instead of the fastest one (number 0). An optimal design of a set of speed profiles can obtain a significant reduction in the energy consumption, removing the requirement for investing in new equipment, as described in [44].

Other scenarios with on-board energy storage devices or network receptivity also benefit from efficient-driving. Reusing regenerative energy from braking to charge storage devices with optimal speed profiles attained 26.77% and 27.36% energy savings in scenarios 4 and 10, respectively, while maintaining a maximum slowdown of 5% in their running time from switching from the fastest speed profile to the designed one.

Overall, due to the traction and braking efficiencies, saving energy from designing optimal speed profiles for efficient-driving is more efficient than maximising the regenerative energy by designing driving patterns that create more braking.

6.4.5 Energy Savings Due to Network Receptivity Improvement and On-Board Energy Storage Devices

In the previous section, Figs. 6.9 and 6.10 have been used to observe the efficiency gained with the nominal driving. However, these figures also show the improvements in energy consumption by improving the network receptivity (for instance, installing track-side storage systems) or on-board energy storage devices when comparing the bars of the different scenarios. Thus, the energy savings obtained by comparing the different scenarios can be calculated. In Table 6.5, the energy savings due to network investment are shown, while the energy savings obtained by investing in on-board storage devices can be seen in Table 6.6.

As a result of the nature of the speed profiles, higher energy savings can be achieved by investing to improve the network receptivity when the flat-out speed profile is considered, rather than with the designed nominal speed, as can be seen in Table 6.5.

Table 6.5 Energy savings due to network investments

	Scenario	Speed profile 0 (%)	Speed profile 1 (%)
Off-peak hours	From 2 to 3	10.19	9.72
	From 5 to 6	1.80	2.07
Peak hours	From 7 to 8	35.64	34.73
	From 8 to 9	2.65	2.6
	From 11 to 12	– 0.45	0.18

Table 6.6 Energy savings due to installing on-board energy storage devices

	Scenario	Speed profile 0 (%)	Speed profile 1 (%)
Off-peak hours	From 2 to 5	7.46	11.67
	From 3 to 6	– 1.18	4.18
Peak hours	From 8 to 11	– 2.50	2.97
	From 9 to 12	– 5.77	0.56

Driving with nominal speed profile in peak hours scenarios where the regenerated energy is transmitted among trains (from scenarios 7 to 8) can save up to 34.73% compared to scenarios with null receptivity. The flat-out speed profiles grant an additional 1% in savings.

Off-peak hours can expect an additional reduction in the total energy consumption of around 10% by installing track-side storage systems (from scenarios 2 to 3). However, this advantage is dramatically reduced in peak hours (from scenarios 8 to 9) because practically all the regenerated energy is used up by the trains. Moreover, when the trains are equipped with on-board energy storage systems, the benefits of improving network receptivity present minimal results (from scenario 5 to 6 and from scenarios 11 to 12).

According to the results shown in Table 6.6, the installation of on-board storage systems is beneficial only in low-density scenarios where no track-side storage system is installed (from scenarios 2 to 5). In other cases, on-board energy storage devices could produce minimal benefits or even an increase in energy consumption, due to the additional train weight.

As a summary, the results show that the investment in energy storage devices is only beneficial in low traffic lines, and does not provide substantial savings in scenarios with high usage of regenerated energy.

6.5 Conclusions

This chapter tackles the application of energy storage systems in railways to maximise the use of regenerated energy and the influence in train efficient-driving. The main technologies applied, the modelling of these energy storage systems and its control has been discussed. Furthermore, the models presented have been applied to a case study using real data from Metro de Madrid. The regenerative energy is considered in the design of efficient-driving evaluating the net energy consumption at substations instead of pantograph by means of a network performance coefficients. The electrical network is characterised by a loss coefficient and an energy recovery coefficient that represents the network receptivity to regenerated energy. Different scenarios are defined to assess energy savings associated to possible investments to improve the use of regenerative energy. These scenarios consider different receptivity values taking into account no network receptivity, a receptivity value that represents the exchange of braking energy among trains and an improved receptivity value representing the installation of track-side energy storage systems. Moreover, the scenarios represent different traffic conditions (high and low density) and the installation or not of on-board energy storage systems.

The results show that significant savings can be achieved by designing optimal speed profiles with associated low investments and the same quality of service. More than 22% of energy savings have been obtained in all the scenarios studied because of efficient-driving application.

Regarding the installation of track-side energy storage systems, it can be concluded that they would provide energy savings in scenarios with low density traffic (between 9.72 and 10.19%), but low energy reduction is obtained in dense traffic scenarios (between 1.8 and 2.07%). On the other hand, installing on-board energy storage systems could be beneficial in low traffic lines with low receptivity to regenerated energy, where savings between 7.46 and 11.67% have been obtained. However, their application in dense traffic conditions or with high network receptivity would increase energy consumption because of the increase in train mass.

References

1. Peña-Alcaraz M, Fernandez A, Cucala AP et al (2011) Optimal underground timetable design based on power flow for maximizing the use of regenerative-braking energy. Proc Inst Mech Eng Part F J Rail Rapid Transit 226:397–408. https://doi.org/10.1177/0954409711429411
2. Greatbanks J (2005) Review of the discount for using regenerative braking. AEA Technol
3. Chymera M, Renfrew A, Barnes M (2008) Analyzing the potential of energy storage on electrified transit systems. In: 8th World congress of railway research–WCRR
4. Kim G, Lee H (2009) A study on the application of ESS on SeoulMetro line 2. In: 2009 international conference on information and multimedia technology, pp 38–42
5. Abrahamsson L, Söder L (2011) Fast estimation of relations between aggregated train power system data and traffic performance. IEEE Trans Veh Technol 60:16–29. https://doi.org/10.1109/TVT.2010.2091293
6. Barrero R, Mierlo JV, Tackoen X (2008) Energy savings in public transport. IEEE Veh Technol Mag 3:26–36. https://doi.org/10.1109/MVT.2008.927485
7. Gunselmann W (2005) Technologies for increased energy efficiency in railway systems. In: 2005 European conference on power electronics and applications, p 10
8. Lee H, Joung E, Kim G, An C (2009) A study on the effects of energy storage system. In: 2009 international conference on information and multimedia technology, pp 28–32
9. Steiner M, Scholten J (2004) Energy storage on board of DC fed railway vehicles PESC 2004 conference in Aachen, Germany, vol 1. In: 2004 IEEE 35th annual power electronics specialists conference (IEEE Cat. No.04CH37551), pp 666–671
10. Adinolfi A, Lamedica R, Modesto C et al (1998) Experimental assessment of energy saving due to trains regenerative braking in an electrified subway line. IEEE Trans Power Deliv 13:1536–1542. https://doi.org/10.1109/61.714859
11. López-López ÁJ, Pecharromán RR, Fernández-Cardador A, Cucala AP (2014) Assessment of energy-saving techniques in direct-current-electrified mass transit systems. Transp Res Part C Emerg Technol 38:85–100. https://doi.org/10.1016/j.trc.2013.10.011
12. Roch-Dupré D, López-López ÁJ, Pecharromán RR et al (2017) Analysis of the demand charge in DC railway systems and reduction of its economic impact with energy storage systems. Int J Electr Power Energy Syst 93:459–467. https://doi.org/10.1016/j.ijepes.2017.06.022

13. Brenna M, Longo M, Yaïci W et al (2017) Modelling and simulation of electric vehicle fast charging stations driven by high speed railway systems. Energies 10:1268. https://doi.org/10.3390/en10091268
14. Falvo MC, Lamedica R, Bartoni R, Maranzano G (2011) Energy management in metro-transit systems: an innovative proposal toward an integrated and sustainable urban mobility system including plug-in electric vehicles. Electr Power Syst Res 81:2127–2138. https://doi.org/10.1016/j.epsr.2011.08.004
15. Fernandez-Rodriguez A, Fernandez-Cardador A, Santiago-Laporte AD et al (2017) Charging electric vehicles using regenerated energy from urban railways. In: 2017 IEEE vehicle power and propulsion conference (VPPC), pp 1–6
16. Fernández-Rodríguez A, Fernández-Cardador A, Cucala AP, Falvo MC (2019) Energy efficiency and integration of urban electrical transport systems: EVs and metro-trains of two real European lines. Energies 12:366. https://doi.org/10.3390/en12030366
17. Gao Z, Fang J, Zhang Y et al (2015) Control of urban rail transit equipped with ground-based supercapacitor for energy saving and reduction of power peak demand. Int J Electr Power Energy Syst 67:439–447. https://doi.org/10.1016/j.ijepes.2014.11.019
18. Iannuzzi D, Ciccarelli F, Lauria D (2012) Stationary ultracapacitors storage device for improving energy saving and voltage profile of light transportation networks. Transp Res Part C Emerg Technol 21:321–337. https://doi.org/10.1016/j.trc.2011.11.002
19. Roch-Dupré D, Gonsalves T, Cucala AP et al (2021) Multi-stage optimization of the installation of energy storage systems in railway electrical infrastructures with nature-inspired optimization algorithms. Eng Appl Artif Intell 104:104370. https://doi.org/10.1016/j.engappai.2021.104370
20. Roch-Dupré D, Camacho-Gómez C, Cucala AP et al (2021) Optimal location and sizing of energy storage systems in DC-electrified railway lines using a coral reefs optimization algorithm with substrate layers. Energies 14:4753. https://doi.org/10.3390/en14164753
21. Arboleya P, Mohamed B, El-Sayed I (2018) DC railway simulation including controllable power electronic and energy storage devices. IEEE Trans Power Syst 33:5319–5329. https://doi.org/10.1109/TPWRS.2018.2801023
22. Guerrieri M (2019) Catenary-free tramway systems: functional and cost-benefit analysis for a metropolitan area. Urban Rail Transit 5:289–309. https://doi.org/10.1007/s40864-019-00118-y
23. Miyatake M, Haga H (2010) Optimization of speed profile and quick charging of a catenary free train with on-board energy storage. In: Railway and ship propulsion electrical systems for aircraft, pp 1–6
24. Hillmansen S, Roberts C (2007) Energy storage devices in hybrid railway vehicles: a kinematic analysis. Proc Inst Mech Eng Part F J Rail Rapid Transit 221:135–143. https://doi.org/10.1243/09544097JRRT99
25. Steiner M, Klohr M, Pagiela S (2007) Energy storage system with ultracaps on board of railway vehicles. In: 2007 European conference on power electronics and applications, pp 1–10
26. Iannuzzi D (2008) Improvement of the energy recovery of traction electrical drives using supercapacitors. In: 2008 13th international power electronics and motion control conference, pp 1469–1474
27. Ogasa M (2008) Energy saving and environmental measures in railway technologies: example with hybrid electric railway vehicles. IEEJ Trans Electr Electron Eng 3:15–20. https://doi.org/10.1002/tee.20227
28. Taguchi Y, Ogasa M, Hata H et al (2007) Simulation results of novel energy storage equipment series-connected to the traction inverter. In: 2007 European conference on power electronics and applications, pp 1–9
29. Lacôte F (2005) Alstom—future trends in railway transportation. Jpn Railw Transp Rev
30. Konishi T, Tobita M (2012) Fixed energy storage technology applied for DC electrified railway (traction power substation). In: Railway and ship propulsion 2012 electrical systems for aircraft, pp 1–6
31. Ratniyomchai T, Hillmansen S, Tricoli P (2014) Recent developments and applications of energy storage devices in electrified railways. IET Electr Syst Transp 4:9–20. https://doi.org/10.1049/iet-est.2013.0031

32. Canova A, Campanelli F, Quercio M (2022) Flywheel energy storage system in italian regional transport railways: a case study. Energies 15:1096. https://doi.org/10.3390/en15031096
33. Gay SE, Ehsani M (2002) On-board electrically peaking drive train for electric railway vehicles, vol 2. In: Proceedings IEEE 56th vehicular technology conference, pp 998–1001
34. Barrero R, Tackoen X, van Mierlo J (2010) Stationary or onboard energy storage systems for energy consumption reduction in a metro network. Proc Inst Mech Eng Part F J Rail Rapid Transit 224:207–225. https://doi.org/10.1243/09544097JRRT322
35. Miyatake M, Haga H, Suzuki S (2009) Optimal speed control of a train with on-board energy storage for minimum energy consumption in catenary free operation. In: 2009 13th European conference on power electronics and applications, pp 1–9
36. Miyatake M, Matsuda K (2009) Energy saving speed and charge/discharge control of a railway vehicle with on-board energy storage by means of an optimization model. IEEJ Trans Electr Electron Eng 4:771–778. https://doi.org/10.1002/tee.20479
37. Agenjos E, Gabaldon A, Franco FG, et al (2009) Energy efficiency in railways: energy storage and electric generation in diesel electric locomotives. In: CIRED 2009—20th international conference and exhibition on electricity distribution—part 1, pp 1–7
38. Arboleya P, El-Sayed I, Mohamed B, Mayet C (2019) Modeling, simulationand analysis of on-board hybrid energy storage systems for railway applications. Energies 12:2199. https://doi.org/10.3390/en12112199
39. Chen J, Hu H, Ge Y et al (2021) An energy storage system for recycling regenerative braking energy in high-speed railway. IEEE Trans Power Deliv 36:320–330. https://doi.org/10.1109/TPWRD.2020.2980018
40. Maher B (2006) Ultracapacitors provide cost and energy savings for public transportation applications. Battery Power Prod Technol 10
41. Brenna M, Bucci V, Falvo MC et al (2020) A review on energy efficiency in three transportation sectors: railways, electrical vehicles and marine. Energies 13:2378. https://doi.org/10.3390/en13092378
42. Cunillera A, Fernández-Rodríguez A, Cucala AP et al (2020) Assessment of the worthwhileness of efficient driving in railway systems with high-receptivity power supplies. Energies 13:1836. https://doi.org/10.3390/en13071836
43. Noda T, Koseki T (2009) Design of a run-curve for energy saving operation in a modern DC-electrification-efficiency assuming perfect regenerative braking, paper-ID 360659. In: International symposium speed-safety service technology railway and Maglev system 2009 STECH09 2009616-19 Niigata Jpn
44. Bocharnikov YV, Tobias AM, Roberts C (2010) Reduction of train and net energy consumption using genetic algorithms for trajectory optimisation. In: IET conference on railway traction systems (RTS 2010), pp 1–5
45. Acikbas S, Soylemez MT (2008) Coasting point optimisation for mass rail transit lines using artificial neural networks and genetic algorithms. IET Electr Power Appl 2:172–182. https://doi.org/10.1049/iet-epa:20070381
46. Khmelnitsky E (2000) On an optimal control problem of train operation. IEEE Trans Autom Control 45:1257–1266. https://doi.org/10.1109/9.867018
47. Su S, Wang X, Cao Y, Yin J (2020) An energy-efficient train operation approach by integrating the metro timetabling and eco-driving. IEEE Trans Intell Transp Syst 21:4252–4268. https://doi.org/10.1109/TITS.2019.2939358
48. Yang X, Ning B, Li X, Tang T (2014) A two-objective timetable optimization model in subway systems. IEEE Trans Intell Transp Syst 15:1913–1921. https://doi.org/10.1109/TITS.2014.2303146
49. Yang X, Li X, Gao Z et al (2013) A cooperative scheduling model for timetable optimization in subway systems. IEEE Trans Intell Transp Syst 14:438–447. https://doi.org/10.1109/TITS.2012.2219620
50. Domínguez M, Fernandez-Cardador A, Cucala AP, Pecharroman RR (2012) Energy savings in metropolitan railway substations through regenerative energy recovery and optimal design of ATO speed profiles. IEEE Trans Autom Sci Eng 9:496–504. https://doi.org/10.1109/TASE.2012.2201148

51. Destraz B, Barrade P, Rufer A, Klohr M (2007) Study and simulation of the energy balance of an urban transportation network. In: 2007 European conference on power electronics and applications, pp 1–10
52. Barrade P, Rufer A (2003) Current capability and power density of supercapacitors: considerations on energy efficiency. In: 10th European conference on power electronics and applications EPE 2003, Toulouse, France, 9
53. Arboleya P, Diaz G, Coto M (2012) Unified AC/DC power flow for traction systems: a new concept. IEEE Trans Veh Technol 61:2421–2430. https://doi.org/10.1109/TVT.2012.2196298
54. Barrero R, Tackoen X, Van Mierlo J (2008) Improving energy efficiency in public transport: stationary supercapacitor based energy storage systems for a metro network. In: 2008 IEEE vehicle power and propulsion conference, pp 1–8
55. Giordano G, Klass V, Behm M et al (2018) Model-based lithium-ion battery resistance estimation from electric vehicle operating data. IEEE Trans Veh Technol 67:3720–3728. https://doi.org/10.1109/TVT.2018.2796723
56. Roch Dupré D (2020) Improving the electrical infrastructure of dc-electrified railway systems to increase energy efficiency, taking into account complex topologies and representative traffic scenarios
57. Fernández-Rodríguez A, Fernández-Cardador A, Cucala AP et al (2015) Design of robust and energy-efficient ATO speed profiles of metropolitan lines considering train load variations and delays. IEEE Trans Intell Transp Syst 16:2061–2071. https://doi.org/10.1109/TITS.2015.2391831
58. Domínguez M, Fernández A, Cucala AP, Lukaszewicz P (2011) Optimal design of metro automatic train operation speed profiles for reducing energy consumption. Proc Inst Mech Eng Part F J Rail Rapid Transit 225:463–474. https://doi.org/10.1177/09544097JRRT420
59. Fernández-Rodríguez A, Cucala AP, Fernández-Cardador A (2020) An eco-driving algorithm for interoperable automatic train operation. Appl Sci 10:7705. https://doi.org/10.3390/app10217705
60. Domínguez M, Fernández-Cardador A, Cucala AP et al (2014) Multi objective particle swarm optimization algorithm for the design of efficient ATO speed profiles in metro lines. Eng Appl Artif Intell 29:43–53. https://doi.org/10.1016/j.engappai.2013.12.015
61. Raquel CR, Naval Jr PC (2005) An effective use of crowding distance in multiobjective particle swarm optimization. In: Proceedings of the 2005 conference on genetic and evolutionary computation. ACM, New York, NY, USA, pp 257–264
62. Rufer A (2010) Energy storage for railway systems, energy recovery and vehicle autonomy in Europe. In: The 2010 international power electronics conference—ECCE ASIA, pp 3124–3127

Chapter 7
Railway Energy Simulation Considering Traction Power Systems

Zhongbei Tian

7.1 Introduction

The electric railway traction system plays a significant role in improving railway energy supply efficiency. To understand the energy consumption performance of the railway operation, it is important to analyze the electrical energy consumption within the traction power network.

The first electric traction power network was built by Werner von Siemens for the Berlin Industrial Exhibition in 1879, which implemented a 150 V DC power network to supply a DC locomotive with 2.2 kW power. With the increasing train power demand and route distance, the modern railway traction power networks are mainly supplied with 600–3000 DC, 15 kV 16 2/3 Hz AC and 25 kV 50 Hz AC voltages. DC electric railway networks account for around a third of electric railway lines across the world. Most modern trains are implemented with regenerative braking systems, which generates electricity during train braking. The traction power network allows the transmission of regenerative braking power from braking trains to motoring trains, which improves the energy efficiency of the railway network.

This chapter will introduce the DC and AC traction network, the modelling methods of railway traction power network and illustrate several case studies to analyze the performance of energy-efficient operation.

Z. Tian (✉)
University of Birmingham, Birmingham, UK
e-mail: z.tian@bham.ac.uk

S. Su et al., *Energy-Efficient Train Operation*, Lecture Notes in Mobility,
https://doi.org/10.1007/978-3-031-34656-9_7

7.2 Railway Traction Power Systems

7.2.1 DC Electric Railway Traction Network

The railway traction power supply system is responsible for providing power energy for vehicles and power supply equipment. The composition of a DC railway traction power supply system is shown in Fig. 7.1 [1], which includes the external distribution power grid and the railway owned internal power supply system. As a user of the distribution power grid, urban rail transit generally obtains electric energy directly from the distribution power grid without building a separate power plant. The internal power supply system of urban rail transit consists of traction power supply system and power lighting power supply system. The traction substation in the traction power supply system converts three-phase high-voltage AC into low-voltage DC suitable for electric vehicle applications. The feeder then sends the DC power of the traction substation to the catenary. The electric vehicle obtains electric energy through the direct contact between its current collector and the catenary. The power lighting power supply system provides power supply for various lighting, escalators, fans, pumps and other power mechanical equipment in stations and sections, as well as power supply for communication, signal, automation, and other equipment. It is composed of a step-down substation and power lighting distribution line.

In modern railways, the DC traction substations are normally equipped with transformers and rectifiers, drawing electricity from local distribution networks. Figure 7.1 presents a typical feeding network for a DC railway system [1]. The electrical supply

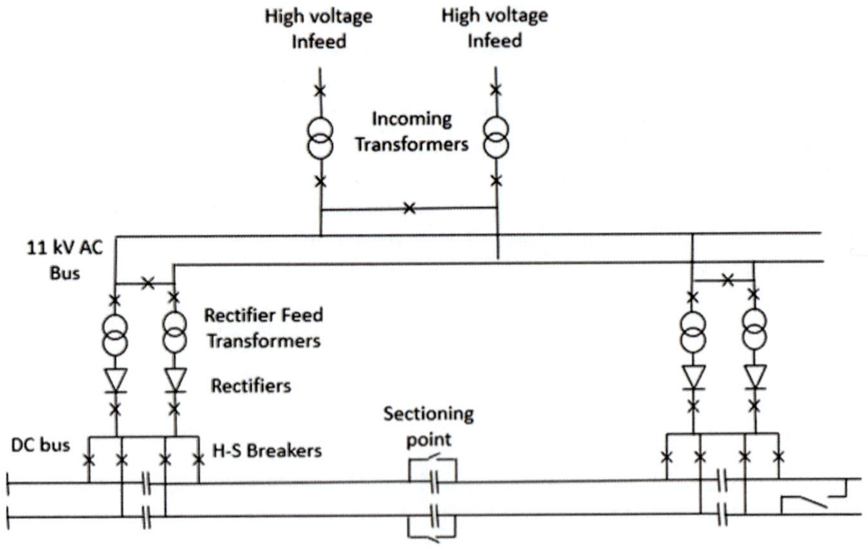

Fig. 7.1 A typical DC feeding arrangement

fed to railways is typically at 132, 66 or 33 kV AC, depending on the size and demand of railway systems. A medium voltage distribution network is normally at 11 kV, fed by step down transformers. The medium voltage network provides energy for the whole railway system. The passenger station is supplied by 415 V 3-phase transformed from 11 kV for domestic usage. Traction substations use transformers and rectifiers to convert 11 kV AC into 600–3000 V DC. The economic distance between substations increases with a higher voltage level. It is recommended to be 4–6 km for 750 V, 8–13 km for 1500 V and 20–30 km for 3000 V [1]. The distance of substations is determined by the power and the number of trains in the network. To improve the efficiency of current transmission, the return rails are normally bonded together and the catenaries are connected at points midway between substations [2].

7.2.2 AC Electric Railway Traction Network

The electrification of an AC railway system is composed of a power supply substation, feeding configuration, overhead catenary system, and trains. Power is supplied from the National Grid, via feeder substations and to the line. The feeder substation is required every 40–60 km [3, 4] with each feeding section separated by a neutral section. The main feeding configurations for AC railway systems are direct feeding, booster transformer, and autotransformer feeding systems [5, 6].

7.2.2.1 Direct Feeding Configuration

The direct feeding system (or Rail-Return), as shown in Fig. 7.2, has 2 conductors: catenary (C) and rail tracks (R) for a single track, and 4 conductors for a double track system (2 conductors and 2 rail tracks). The voltage level of the direct feeding system is 25 kV, where the potentials at conductors C and R are 25 kV and 0 V, respectively. A substation of 25 kV, represented by a voltage source, is connected to the catenary and rail tracks. The traction current from the substation is supplied via the overhead catenary wire through the pantograph, and the return current passes through the rail tracks as return conductor and returns to the substation.

This is a simple and cheap method of electrification, however, there are drawbacks; high impedance which causes high power loss, high rail-to-earth voltage and high interference to lineside signalling and communications. To reduce the interference due to the inductive coupling from the electrification system, the separation between

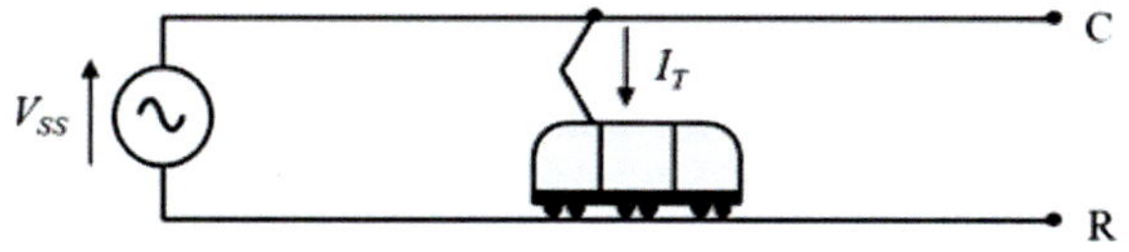

Fig. 7.2 Direct feeding configuration

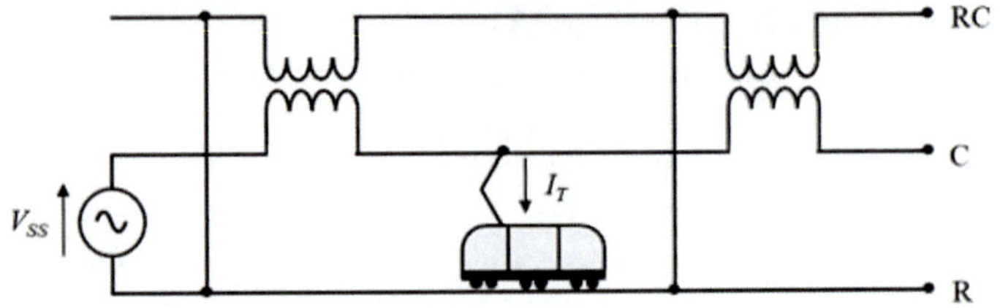

Fig. 7.3 Booster transformer configuration

the overhead transmission line and the return conductor needs to be reduced [5, 6]. Therefore, the booster transformer configuration was introduced.

7.2.2.2 Booster Transformer Configuration

The booster transformer (BT) configuration as shown in Fig. 7.3, requires a BT every 3–4 km along the line. This configuration has 3 conductors: return conductor (RC), catenary (C) and rail tracks (R) for a single track, and 6 conductors for double track system (2 return conductors, 2 catenaries and 2 rail tracks). The voltage level of the BT system is 25 kV, where the potentials at conductors RC, C and R are 0 V, 25 kV and 0 V, respectively. A substation of 25 kV, represented by a voltage source, is connected to the catenary and rail tracks. A parallel post connects the return conductors and rail tracks for the return current. The train pulls current from the overhead catenary wire, passes through the rail tracks and returns via the return conductor.

The overhead return conductor reduces the magnetic field interferences to the lineside signalling and communications. However, this system has a high impedance along the overhead transmission line, which results in higher voltage drops and higher power losses [5, 6].

7.2.2.3 Autotransformer Configuration

For the autotransformer (AT) configuration, as shown in Fig. 7.4, the AT is required every 8–15 km along the railway line. This configuration has 3 conductor wires: catenary (C), rail tracks (R) and feeder wire (F) for a single track, and 6 conductors for double track system (2 catenaries, 2 rail tracks and 2 feeder wires). The voltage level of the AT system is 2×25 kV, which reduces the current supplied by the substation to a half of the train current. The potentials at conductors C, R and F are 25 kV, 0 V and −25 kV, respectively. A substation of 2×25 kV, represented by a voltage source, is connected to the catenary and feeder wires. A train draws current from the catenary, passing through the rail tracks, the autotransformer and returns via the feeder wire.

There are various benefits such as reducing the number of feeder stations; the interference is reduced greatly from using feeder lines and opposite current directions.

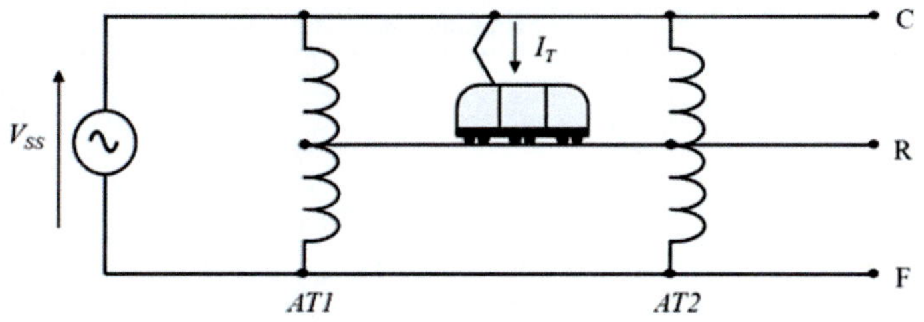

Fig. 7.4 Autotransformer configuration

The autotransformer feeding system uses a higher voltage at 50 kV which reduces current and therefore has lower loss in the transmission line. It has higher power transmission and needs fewer substations, which helps reduce the infrastructure cost. However, this feeding system requires a larger AT sizing, which is more expensive than the BT [5, 6].

7.3 Mathematical Modelling of Railway Traction Power Systems

7.3.1 DC Traction Substation

In a DC railway power network, the traction rectifier substations are the primary electricity source for vehicles. Figure 7.5 presents a typical DC traction power network with multiple trains on up and down tracks. The rectifier substation is connected to the DC busbar, which feeds the power network in both the up and down tracks. When the transmission line voltage is higher than the substation voltage itself, the rectifier substation will prevent current from flowing back to the AC utility grid. This section introduces the method to simulate the components in the power network by equivalent electric circuits.

Overall, the electrical supply substation is equipped with three phase 6-pulse or 12-pulse rectifiers, as shown in Fig. 7.6. With the development of power electronic techniques, equivalent 24-pulse rectifiers are being applied in modern rapid transit systems, where two 12-pulse rectifiers are combined in parallel. The voltage regulation characteristic of the rectifier units is nonlinear, where the ratio of output voltage to current depends on the loads [7]. In order to simplify the simulation analysis, this study limits the working region of the rectifier units. Only the voltage regulation characteristic at normal loads is considered in this book. Thus, the voltage regulation characteristic can be simplified as linear. For example, in Fig. 7.6, the no-load voltage (850 V) decreases linearly with the current. The rated voltage and current are 750 V and 2500 A, respectively. The equivalent resistance for this rectifier substation can be calculated by Eq. (7.1).

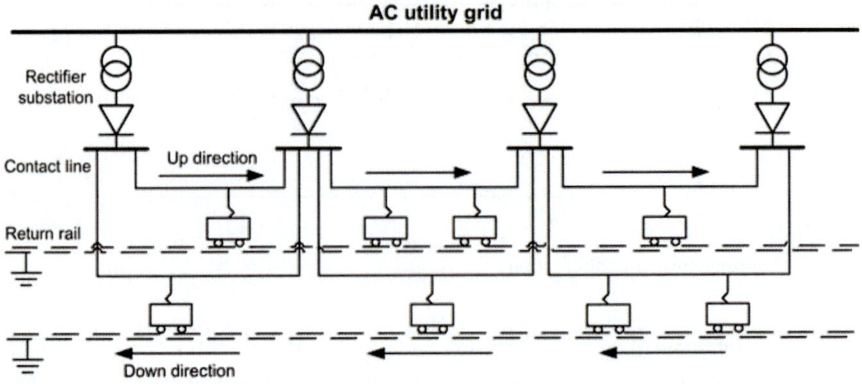

Fig. 7.5 Typical DC traction power network

$$R_{sub} = \frac{\Delta V}{\Delta I} = \frac{V_{noload} - V_{rated}}{I_{rated} - 0} = \frac{850 - 750}{2500 - 0} = 0.04\ \Omega \tag{7.1}$$

A rectifier substation can be modelled by an ideal voltage source in series with an equivalent source resistance and a diode, as in Fig. 7.7. The voltage source is equal to the no-load voltage of the substation.

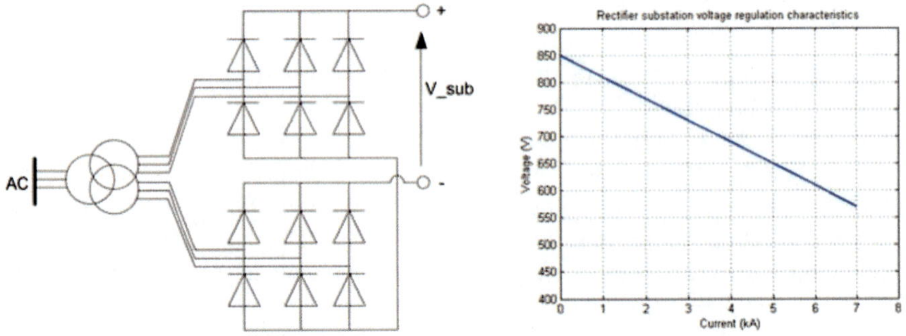

Fig. 7.6 A 12-pulse rectifier unit and voltage regulation characteristics

Fig. 7.7 Equivalent circuit of rectifier substation

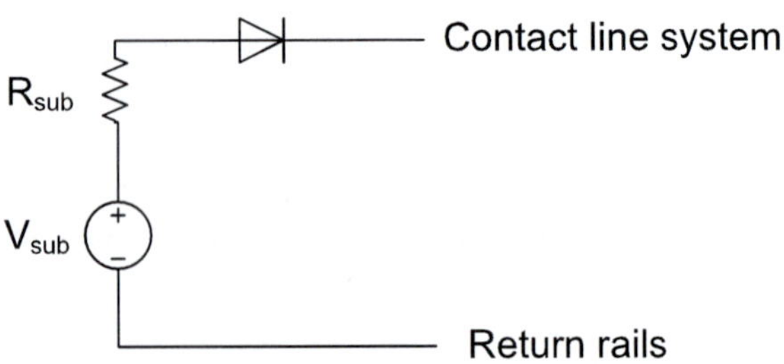

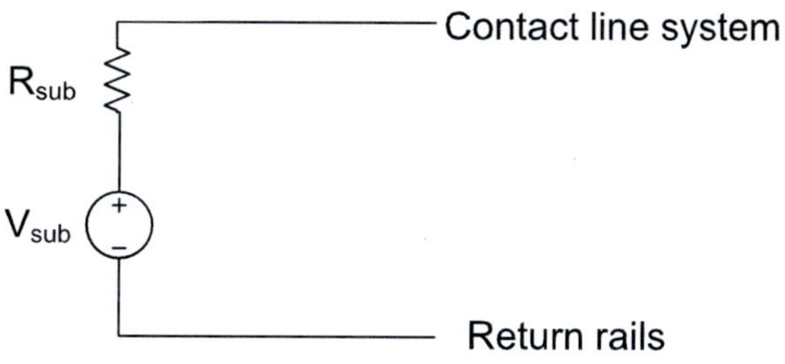

Fig. 7.8 Equivalent circuit of rectifier substation switched on

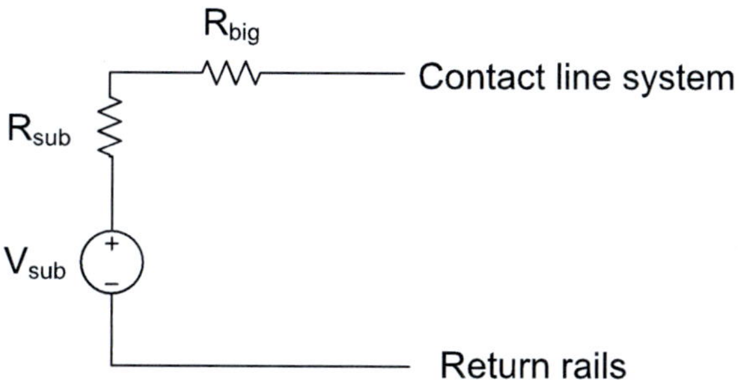

Fig. 7.9 Equivalent circuit of rectifier substation switched off

In practice, there are two working modes for a rectifier substation. If the contact line voltage is lower than the no-load voltage, the diode in Fig. 7.7 is forward biased and the substation delivers power to the network. In this case, the substation equivalent circuit can be represented as Fig. 7.8.

If the contact line voltage is higher than the no-load voltage, the diode in Fig. 7.7 is reverse-biased and the substation does not deliver power to the network. The equivalent circuit of the substation is presented in Fig. 7.9, where the voltage source is in series with an inner resistor and a very large resistance. The very large resistance is 10^6 Ω in this simulation [8].

In order to simplify the power flow analysis, the voltage source of the substation model is transformed into a current source using Thevenin's and Norton's theorem [9], as shown in Fig. 7.10. The substation current can be calculated by Eq. (7.2), where 'u' is equal to 0 when the rectifier substation is conducting, and otherwise is equal to 1.

$$I_{sub} = \frac{V_{sub}}{R_{sub} + u \times R_{big}} \tag{7.2}$$

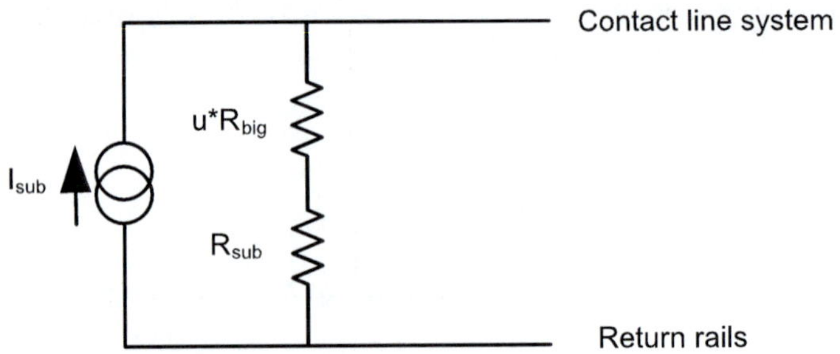

Fig. 7.10 Current source circuit of substation model

7.3.2 AC Traction Substation

In an AC railway power network, the traction substations are the primary electricity source for vehicles. Figure 7.5 also represents a typical AC traction power network with multiple trains on up and down tracks. However, the substation is connected to the AC busbar instead, which feeds the power network in both the up and down tracks. This section introduces the method to simulate the components in the power network by equivalent electric circuits.

An AC substation can be modelled by an ideal voltage source in series with an equivalent transformer impedance, as in Fig. 7.11. The voltage source is equal to the no-load voltage of the substation, which is 27.5 kV for AC traction substation.

In order to simplify the power flow analysis, the voltage source of the AC substation model is transformed into a current source using Thevenin's and Norton's theorem, as shown in Fig. 7.12. The substation current can be calculated by Eq. (7.3).

$$I_{sub} = \frac{V_{sub}}{Z_{sub}} \tag{7.3}$$

Fig. 7.11 Equivalent circuit of AC substation

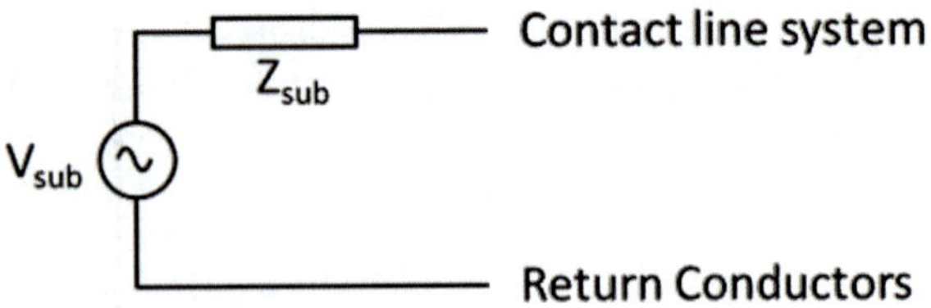

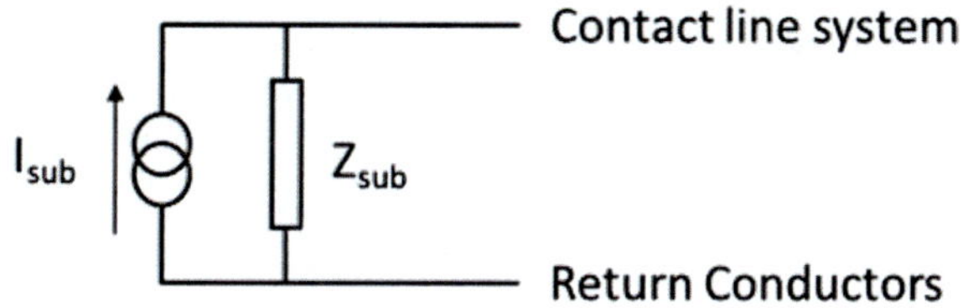

Fig. 7.12 Current source circuit of AC substation model

7.3.3 Dynamic Train Loads

Some previous research used constant current source models or constant efficiency of regenerative braking energy usage to present trains in a traction power network [10–13]. However, this is not accurate in the study of energy consumption for railways. In railway power systems, modern trains collect electricity behaving as voltage-dependant power loads. The power consumed by trains does not depend on the voltage or current at the pantograph [14]. In this chapter, trains are considered as dynamical power sources or power loads for a better simulation performance.

The mechanical power required by the train is determined by the driving operation controls, which has been explained in the previous chapters. The mechanical power for each instant time can be simply calculated in Eq. (7.4).

$$P_{me} = F \times v \tag{7.4}$$

where:

- F is the tractive effort (N);
- v is the vehicle speed (m/s).

In order to analyse the power flow in the power network simulation, the electrical power requirements can be transformed from the mechanical power results as in Eq. (7.5). The efficiency (η) refers to the whole traction chain from the current collector to the wheel, which is around 85%. The positive mechanical power will lead to a higher positive electrical power requirement. The negative mechanical power is the braking power, which will lead to a lower electric regenerative braking power.

$$P_{el} = \begin{cases} \frac{P_{me}}{\eta} & \text{if } P_{me} \geq 0 \\ P_{me} \times \eta & \text{if } P_{me} < 0 \end{cases} \tag{7.5}$$

where:

- P_{me} is the train mechanical power (W);
- P_{el} is the train electrical power (W).

Train electrical power is positive when the train is motoring and becomes negative when the train is braking. The auxiliary load of the train is the power used by the air conditioning and lighting power, which is assumed as a constant. The vehicle power demand (P_{train_demand}) is computed by summing the auxiliary load power and electrical power of the motor, given in Eq. (7.6). This value is treated as an input to the power network simulator.

$$P_{train\_demand} = P_{el} + P_{aux} \tag{7.6}$$

where:

- P_{train_demand} is the train electrical power demand (W);
- P_{aux} is the train auxiliary load power (W).

The power network simulator will solve the power flow. The train voltage and current can be calculated. P_{train} is the electric power which the traction train receives or the braking train exports, given as

$$P_{train} = I_{train} \times V_{train} \tag{7.7}$$

where:

- P_{train} is the train received electrical power (W);
- I_{train} is the train electrical current (A);
- V_{train} is the train electrical voltage (V).

If the train is running in a normal mode, the final train electric power is equal to the train power demand. If not, the train may receive less power compared with the demand, while the braking train may not export all of the electric braking power back to the network. The train modelling methods for traction and braking modes are illustrated as follows.

7.3.3.1 Traction Train

When the train is motoring, it collects power from the power network, transforming the input electrical power into mechanical power. In order to protect the train and supply networks, the train is equipped with automatic devices which are able to adapt the power demand. From the British Standard in Railway applications-Power supply and rolling stock [15], the maximum allowable train current against the train voltage is given in Fig. 7.13.

According to the pantograph voltage, the working state of trains is categorised into three zones. In zone 1 where the train voltage is lower than V_{min2}, there is no traction supply from the power network. The train only collects the electricity to feed the auxiliary system. In zone 2, the train is operated in under-voltage traction mode. The train power is limited, even though the train requires higher power. In zone 3, the train is operated in normal traction. V_n is the nominal voltage of the system, and a is the knee point factor which is lower than 1 (normally between 0.8 and 0.9). When the train voltage is higher than $a \times V_n$, the train can be supplied with maximum traction power. Taken from the British Standard in Railway Applications—Supply

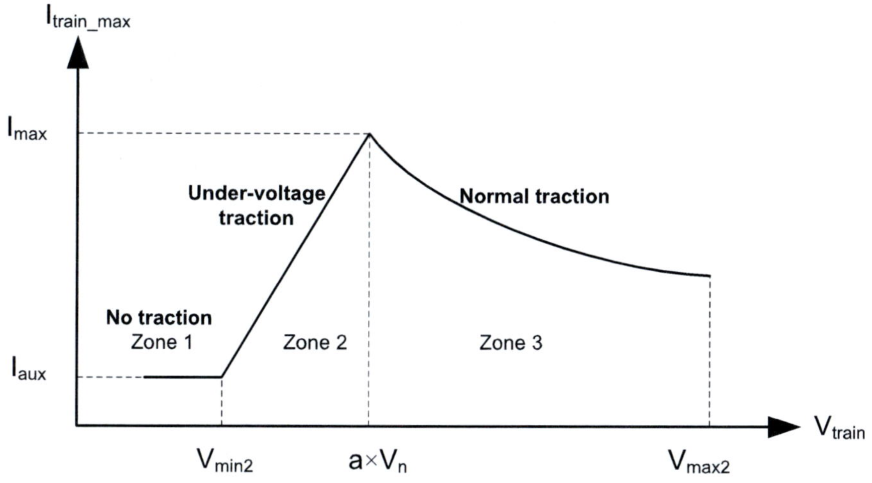

Fig. 7.13 Current limitation of a traction train

voltages of traction systems [16], the voltage characteristics including under-voltage and over-voltage levels for DC railway systems are specified in Table 7.1.

For traction trains, the maximum train power demand is the sum of maximum traction power and auxiliary power.

$$P_{train\_demand\_max} = \frac{P_{trac\_me\_max}}{\eta} + P_{aux} \tag{7.8}$$

where:

- $P_{train\_demand\_max}$ is the maximum train electrical power demand (W);
- $P_{trac\_me\_max}$ is the maximum mechanical traction power (W);
- η is the efficiency of traction chain conversion.

Table 7.1 Voltage permissible limits for DC railways [16]

Traction power systems	Lowest non-permanent voltage V_{min2} (V)	Lowest permanent voltage V_{min1} (V)	Nominal voltage V_n (V)	Highest permanent voltage V_{max1} (V)	Highest non-permanent voltage V_{max2} (V)
600 V DC	400	400	600	720	800
750 V DC	500	500	750	900	1000
1500 V DC	1000	1000	1500	1800	1950
3000 V DC	2000	2000	3000	3600	3900
15 kV AC	11,000	12,000	15,000	17,250	18,000
25 kV AC	17,500	19,000	25,000	27,500	29,000

According to the current limitation applied to the current at the current collector shown in Fig. 7.13, the maximum current and the auxiliary current of the train are calculated by

$$I_{max} = \frac{P_{train\_demand\_max}}{a \times V_n} \quad (7.9)$$

$$I_{aux} = \frac{P_{aux}}{V_{min2}} \quad (7.10)$$

If the train voltage is known, the maximum train traction current can be expressed by V_{train} using

$$I_{train\_max} = \begin{cases} I_{aux} & \text{if } V_{train} \leq V_{min2} \\ \frac{V_{train} - V_{min2}}{r_{trac\_eq}} + I_{aux} & \text{if } V_{min2} < V_{train} \leq a \times V_n \\ \frac{P_{train\_demand\_max}}{V_{train}} & \text{if } V_{train} > a \times V_n \end{cases} \quad (7.11)$$

where the equivalent train under-voltage traction equivalent resistance can be calculated as

$$r_{trac\_eq} = \frac{a \times V_n - V_{min2}}{I_{max} - I_{aux}} \quad (7.12)$$

At different train voltage levels, the maximum electrical power which the train can collect from the power network is calculated by train current and voltage, as in Eq. (7.13). It can be also defined in different zones according to the train voltage values in Eq. (7.14). The maximum received power curve is shown in Fig. 7.14. When the train is operated in no traction zone 1, it is assumed that only the auxiliary power can be supplied. When the train is operated in under-voltage traction mode, the train cannot be supplied with enough power when it requires higher power than the limits. In zone 3, the train can be provided with enough power even when it requires the maximum traction power.

$$P_{train\_max} = I_{train\_max} \times V_{train} \quad (7.13)$$

$$P_{train\_max} = \begin{cases} P_{aux} & if\ V_{train} \leq V_{min2} \\ \frac{(V_{train} - V_{min2}) \times V_{train}}{r_{trac\_eq}} + P_{aux} & if\ V_{min2} < V_{train} \leq a \times V_n \\ P_{train\_demand\_max} & if\ V_{train} > a \times V_n \end{cases} \quad (7.14)$$

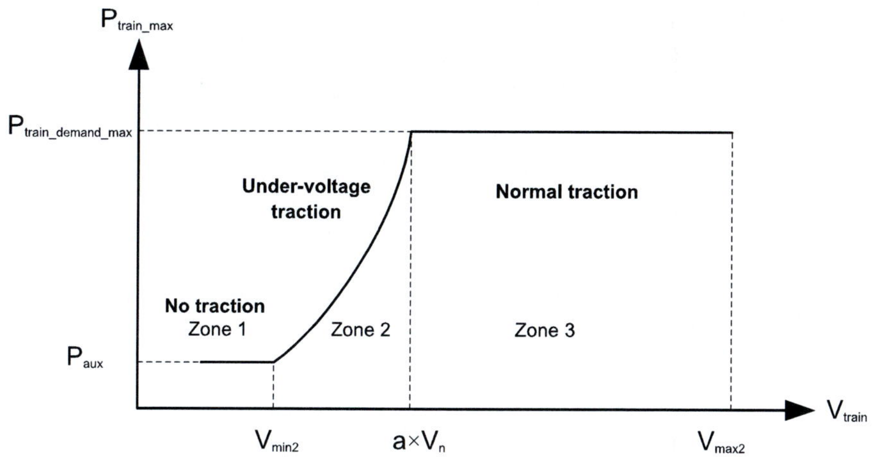

Fig. 7.14 Received power limitation of traction train

7.3.3.2 Traction Train Equivalent Circuit

In normal traction mode, the equivalent circuit of the train is shown in Fig. 7.15. The train is modelled as a dynamic power load. The received train power is equal to the train power demand. The relation between the train power demand and train power received, as well as the train current and voltage can be expressed by

$$P_{train\_demand} = P_{train} = I_{train} \times V_{train} \tag{7.15}$$

In under-voltage traction mode, the train can be modelled as Fig. 7.16. The train received power is lower than the train power demand, as illustrated in Eq. (7.16). The train current can be expressed using the train voltage in Eq. (7.17). It is the sum

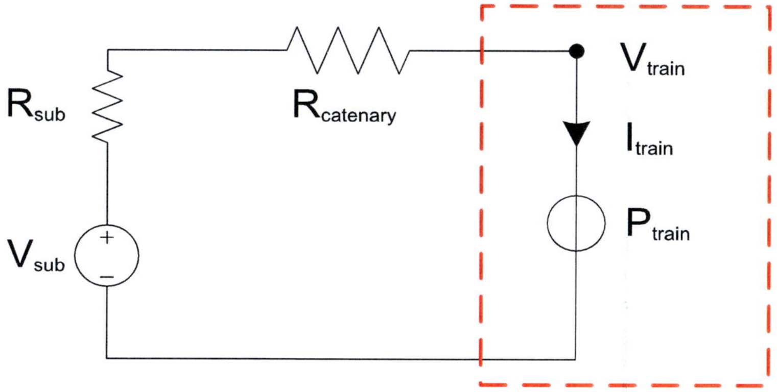

Fig. 7.15 Equivalent circuit of traction train in normal operation

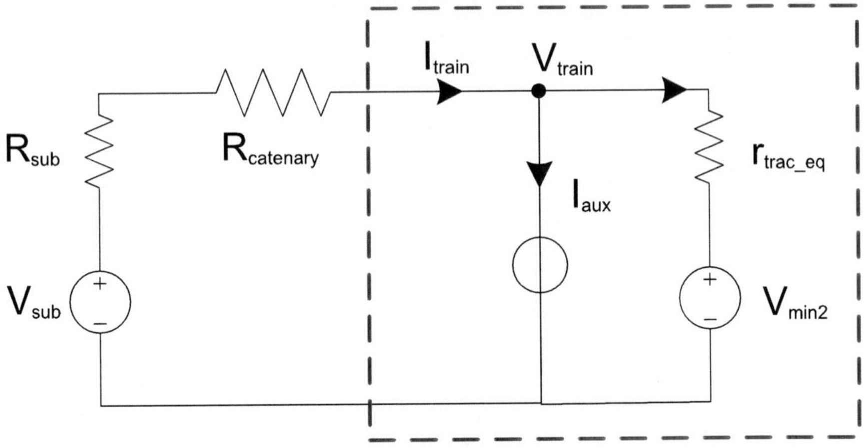

Fig. 7.16 Equivalent circuit of traction train in under-voltage

of the constant train auxiliary current and dynamic traction current which depends on the train voltage.

$$P_{train\_demand} > P_{train} = I_{train} \times V_{train} \tag{7.16}$$

$$I_{train} = I_{aux} + \frac{V_{train} - V_{min2}}{r_{trac\_eq}} \tag{7.17}$$

In the non-traction mode, the train current can be assumed as a constant current source. In this abnormal situation, the train receives no traction power. The train can only be operated in this condition for a short period [17].

7.3.3.3 Braking Train

When the train is in regenerative braking mode, the motor will transform the mechanical power of the drive shaft into electrical power. The regenerated power can be used by the on-board auxiliary system and the surplus regenerated power can be transferred back to the network system feeding other motoring trains. In normal regenerative braking mode, all of the regenerative braking power can be transferred back to the network. However, the regenerative braking can increase train voltage, and a high regen voltage will occur when there are not enough motoring trains absorbing the regenerative energy in the power network. In the case of a high voltage hazard, some of the braking energy cannot be transferred to contact lines. Instead, the energy is wasted in the on-board braking rheostat as heat, when the regen voltage is in excess of a safe value. In the overvoltage regenerating mode, the train current will be limited automatically by the electronic devices. The maximum allowable train

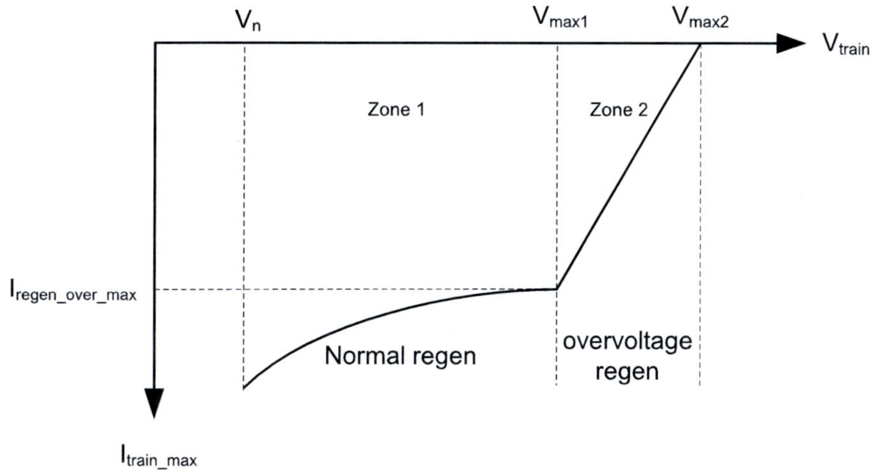

Fig. 7.17 Current limitation of braking train

current against the train voltage is given in Fig. 7.17 [15]. The braking train can be operated in normal regeneration mode when the train voltage is lower than V_{max1}, which is specified in Table 7.1. Normally, the braking train voltage is higher than V_n. When the train voltage exceeds V_{max1}, the train will be operated in over-voltage mode. The maximum train current is limited and becomes zero when the train voltage reaches V_{max2}.

Regarding the braking train, the maximum train power that can be transferred into the power network is the maximum electric braking power plus auxiliary power.

$$P_{train\_demand\_max} = P_{brake\_me\_max} \times \eta + P_{aux} \tag{7.18}$$

where:

- $P_{brake\_me\_max}$ is the maximum mechanical braking power (W).

According to Fig. 7.17, in over-voltage regeneration mode, the train current at V_{max1} is given as

$$I_{regen\_over\_max} = \frac{P_{train\_demand\_max}}{V_{max1}} \tag{7.19}$$

If the train voltage is known, the maximum braking train current can be expressed by V_{train} as

$$I_{train\_max} = \begin{cases} \frac{P_{train\_demand\_max}}{V_{train}} & \text{if } V_{train} \leq V_{max1} \\ \frac{V_{train} - V_{max2}}{r_{brake\_eq}} & \text{if } V_{max1} < V_{train} \leq V_{max2} \end{cases} \tag{7.20}$$

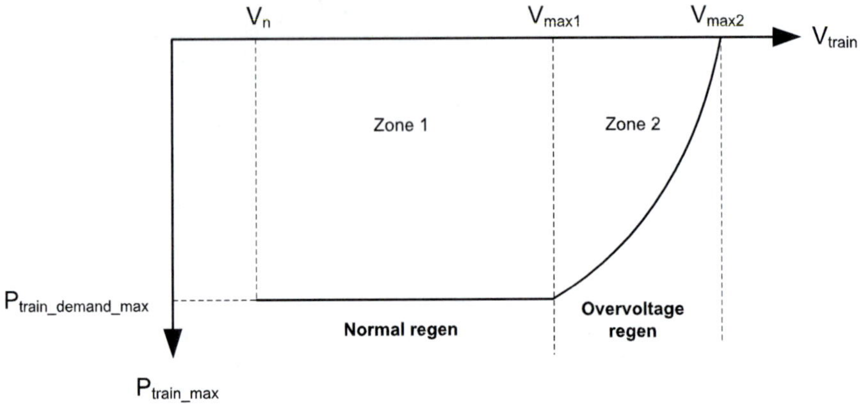

Fig. 7.18 Exported power limitation of braking train

The equivalent train over-voltage braking equivalent resistance can be calculated by

$$r_{brake\_eq} = \frac{V_{max1} - V_{max2}}{I_{regen\_over\_max}} \tag{7.21}$$

At different train voltage levels, the maximum electrical power which the train can transfer into the power network is given in Eq. (7.22), which can be also calculated by Eq. (7.23). The maximum exported power curve is described in Fig. 7.18. When the train is operated in normal regeneration zone 1, all of the electrical braking power can be exported. When the train is operated in overvoltage regeneration mode, the train can only export part of the total electrical braking power. When train voltage is equal to V_{max2}, no electrical braking power can be transferred into the power network.

$$P_{train\_max} = I_{train\_max} \times V_{train} \tag{7.22}$$

$$P_{train\_max} = \begin{cases} P_{train\_demand\_max} & if\ V_{train} \le V_{max1} \\ \frac{(V_{train} - V_{max2}) \times V_{train}}{r_{brake\_eq}} & if\ V_{max1} < V_{train} \le V_{max2} \end{cases} \tag{7.23}$$

7.3.3.4 Braking Train Equivalent Circuit

In normal regeneration mode, the equivalent circuit of the train is shown in Fig. 7.19. The train is modelled as a dynamic power load. The train exported power is equal to the train power demand. The relation between the train power demand, train power received and the train current and voltage can be expressed in Eq. (7.24).

$$P_{train\_demand} = P_{train} = I_{train} \times V_{train} \tag{7.24}$$

In overvoltage regeneration mode, the train is modelled as in Fig. 7.20. The train exported power is lower than the train power demand as in Eq. (7.25). The train current can be expressed using the train voltage as shown in Eq. (7.26).

$$\left|P_{train\_demand}\right| > \left|P_{train}\right| = \left|I_{train} \times V_{train}\right| \tag{7.25}$$

$$I_{train} = \frac{V_{train} - V_{max2}}{r_{brake\_eq}} \tag{7.26}$$

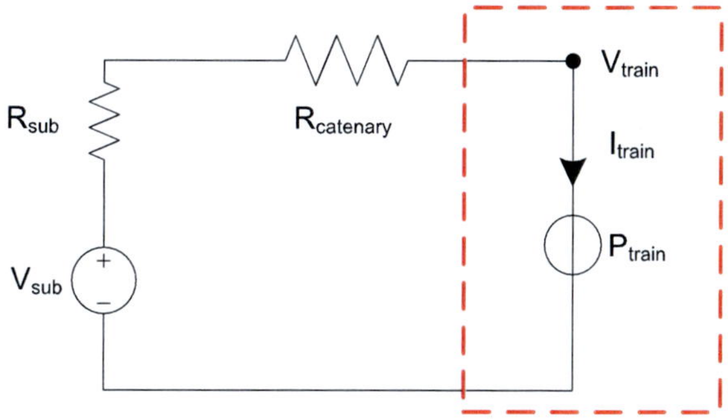

Fig. 7.19 Equivalent circuit of braking train in normal operation

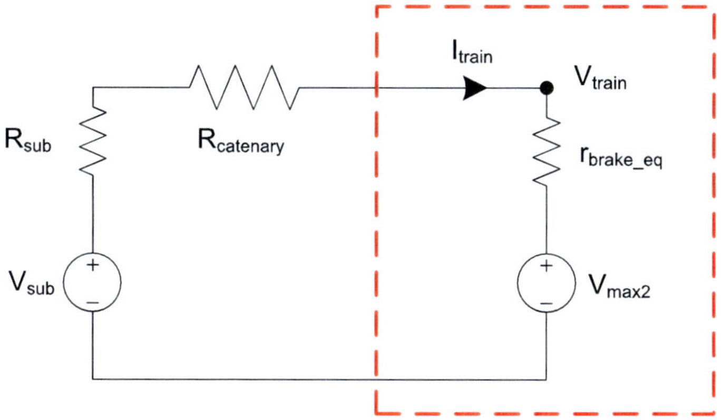

Fig. 7.20 Equivalent circuit of braking train in over-voltage

7.3.4 Admittance Matrix Construction

An example of the DC railway system equivalent circuit is described in Fig. 7.21, which consists of the equivalent models of substations and trains as explained in previous sections. The paralleling post connects the contact lines on both tracks for reducing transmission losses as well as improving line voltages. It is modelled by a zero-resistance conductor connecting two contact lines. The substation connects with the busbar to feed both tracks. In practice, the rails bond together every 250 or 500 m. Two tracks can be modelled by one combined return rail for admittance simplification with a reasonably low error [18]. The conductor resistors are used to represent the overhead line and return running rail resistance, which are split by trains, substations and parallel posts. The resistance of the contact line and lumped rail depends on the length and resistivity of the conductor as in Eq. (7.27), where L refers to the length of the conductor, ρ_c and ρ_r refer to the resistivity of the contact line and return rail per track. The resistivity of overhead conductor systems is in the range of 30 to 90 mΩ/km, whereas it is between 8 and 20 mΩ/km for the third rail [2, 19]. The resistivity of the return rail is around 20 mΩ/km/track [17].

$$\begin{cases} R_c = L \times \rho_c \\ R_r = L \times \frac{\rho_r}{2} \end{cases} \tag{7.27}$$

The admittance matrix of a railway power network circuit is complex. With the chain circuit topology of railway equivalent circuit, the admittance matrix of each component can be constructed and connected conveniently [20]. Figure 7.22

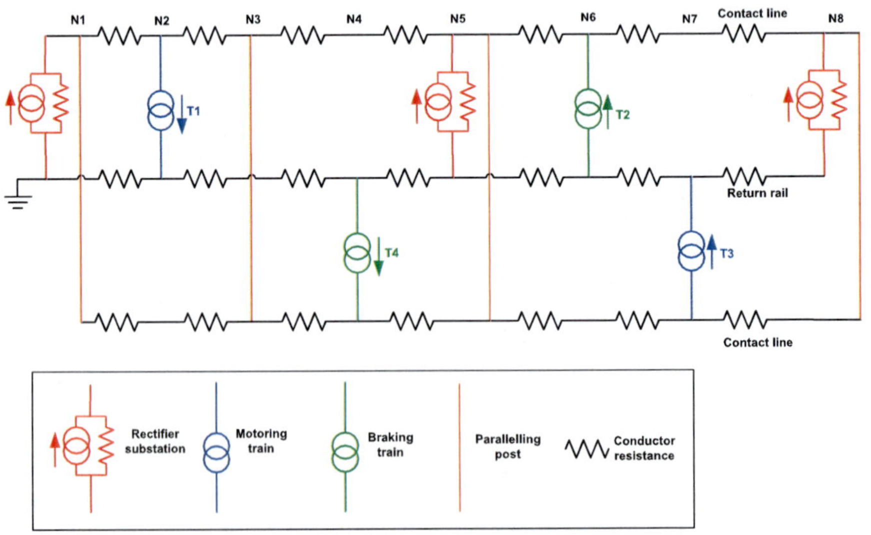

Fig. 7.21 An example of railway system equivalent circuit

describes a chain circuit topology of a railway equivalent circuit with 3 paralleling layers. The circuit is classified by N − 1 serial conductors and N shunt sections. The serial conductors represent the split resistances of contact lines and lumped rails. The shunt sections include the parallel posts, substations and trains, which can separate the conductor lines.

The nodal analysis equation can be applied to solve the power flow of the railway network in Eq. (7.28). The admittance matrix of the whole network (Y) is a 3N × 3N matrix, while both the current and voltage vector is a 1 × 3N matrix. Matrix Y is composed of the admittance matrix of each serial conductor element (Y_c) and shunt element (Y_s), which are 3 × 3 matrices. The admittance matrix of the whole network is expressed in Eq. (7.29), which is a sparse matrix.

$$I = Y \times V \tag{7.28}$$

$$Y = \begin{bmatrix} Y_{s1} + Y_{c1} & -Y_{c1} & 0 & & & \\ -Y_{c1} & Y_{c1} + Y_{s2} + Y_{c2} & -Y_{c2} & \cdots & 0 & \\ 0 & -Y_{c2} & Y_{c2} + Y_{s3} + Y_{c3} & & & \\ & \vdots & & \ddots & \vdots & \\ & 0 & & \cdots & Y_{cN-2} + Y_{sN-1} + Y_{cN-1} & -Y_{cN-1} \\ & & & & -Y_{cN-1} & Y_{cN-1} + Y_{sN} \end{bmatrix} \tag{7.29}$$

The current and voltage vectors in Eq. (7.28) are both 1 × 3N matrices, which are composed of nodal voltage vector (V_s) and current vectors (I_s) of each shunt section, as in Eqs. (7.30) and (7.31). Both V_s and I_s are 1 × 3 matrices. According to the features of the railway power network, the serial and shunt elements can be concluded by the following forms.

$$V = \left[V_{S1} V_{S2} V_{S3} \ldots V_{SN-1} V_{SN}\right]^T \tag{7.30}$$

$$I = \left[I_{S1} I_{S2} I_{S3} \ldots I_{SN-1} I_{SN}\right]^T \tag{7.31}$$

The abovementioned DC railway is modelled as a lumped rail model, with 3 conductors: two contact lines and lumped rails. However, the AC railway is not

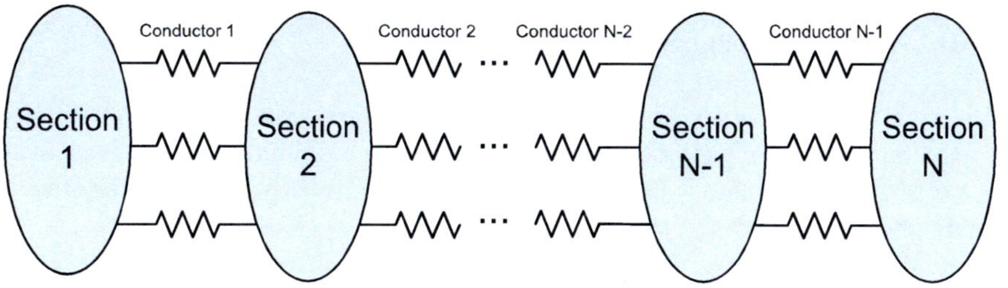

Fig. 7.22 Chain circuit topology of railway equivalent circuit

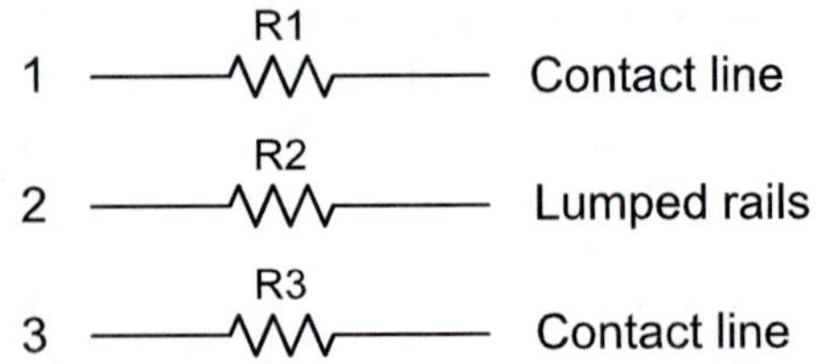

Fig. 7.23 Serial conductor element

modelled with lumped conductors. An AC railway model with direct feeding configuration consists of 4 conductors: two catenary wires (C) and two rails (R), therefore, the admittance matrix of the whole network (Y) is a 4N × 4N matrix, while both the current and voltage vector is a 1 × 4N matrix. Matrix Y is composed of the admittance matrix of each serial conductor element (Y_c) and shunt element (Y_s), which are 4 × 4 matrices. For BT and AT configurations, they consist of 6 conductors: two catenary wires (C), two return conductors (RC) and two rails (R). While an additional return conductor (RC) is used in the BT system, a feeder wire (F) is used in the AT system. Therefore, the admittance matrix of the whole network (Y) is a 6N × 6N matrix, while both the current and voltage vector is a 1 × 6N matrix. Matrix Y is composed of the admittance matrix of each serial conductor element (Y_c) and shunt element (Y_s), which are 6 × 6 matrices.

7.3.4.1 Serial Conductor Elements

The serial conductors consist of the contact lines and the lumped rails. The resistance can be calculated by Eq. (7.27). One section of serial conductors is shown in Fig. 7.23. The admittance matrix of this part can be expressed in Eq. (7.32).

$$Y_c = \begin{bmatrix} \frac{1}{R_1} & 0 & 0 \\ 0 & \frac{1}{R_2} & 0 \\ 0 & 0 & \frac{1}{R_3} \end{bmatrix} \tag{7.32}$$

7.3.4.2 Shunt Impedance

The shunt resistance is a basic model for the shunt section, which connects paralleling conductor lines. An example is shown in Fig. 7.24. The admittance for this element is expressed in Eq. (7.33). The self-admittances (Y_{s11} and Y_{s22}) and the mutual admittances (Y_{s12} and Y_{s21}) are equal to $\frac{1}{R}$.

$$Y_s = \begin{bmatrix} \frac{1}{R} & -\frac{1}{R} & 0 \\ -\frac{1}{R} & \frac{1}{R} & 0 \\ 0 & 0 & 0 \end{bmatrix} \tag{7.33}$$

Fig. 7.24 Shunt resistance

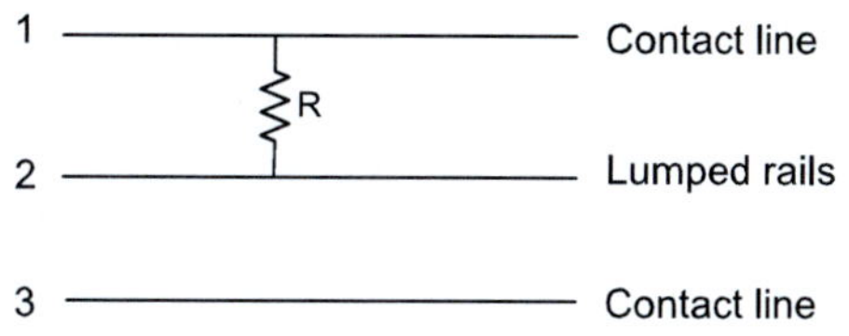

The parallel post which connects two contact lines is shown in Fig. 7.25. According to the admittance matrix structure of a shunt resistance, the parallel post can be assumed as a very small resistor connecting both contact lines. Therefore, the admittance matrix of the parallel post can be expressed in Eq. (7.34), where the small resistance (R_{small}) is set to 10^{-6} Ω in this book [8].

$$Y_s = \begin{bmatrix} \frac{1}{R_{small}} & 0 & -\frac{1}{R_{small}} \\ 0 & 0 & 0 \\ -\frac{1}{R_{small}} & 0 & \frac{1}{R_{small}} \end{bmatrix} \tag{7.34}$$

Similarly, for a grounding connection in Fig. 7.26, the admittance matrix can be written as in Eq. (7.35). The self-admittance Y_{s22} is equal to $\frac{1}{R_{small}}$, while all the mutual admittances are equal to zero.

$$Y_s = \begin{bmatrix} 0 & 0 & 0 \\ 0 & \frac{1}{R_{small}} & 0 \\ 0 & 0 & 0 \end{bmatrix} \tag{7.35}$$

Fig. 7.25 Parallel post

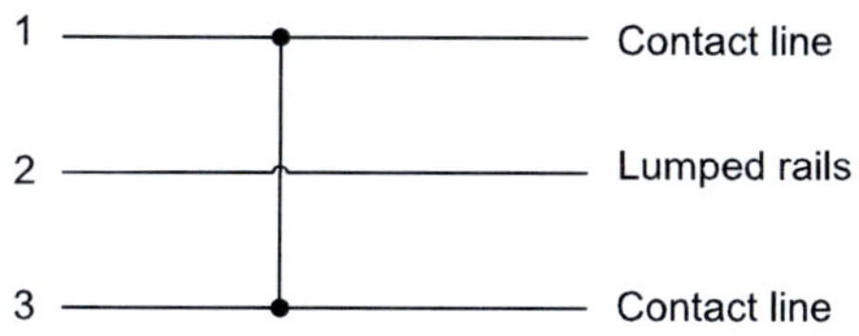

Fig. 7.26 Grounding connection

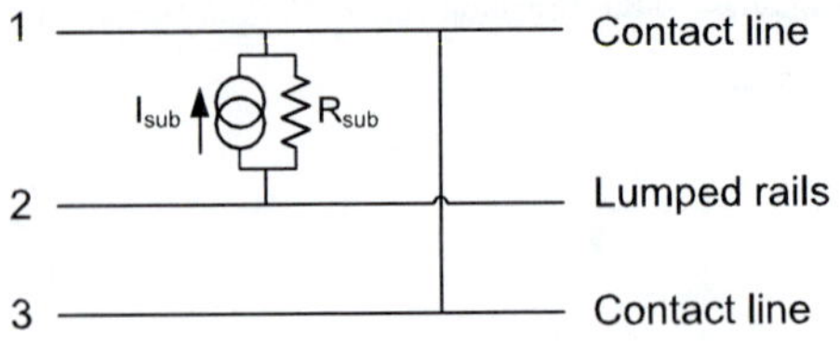

Fig. 7.27 Substation element

7.3.4.3 Substations

The substation is composed of a current source, a resistance and parallel post. The ideal current source does not affect the admittance matrix. The admittance matrix of a substation can be calculated by Eq. (7.36), which is the sum of the admittance matrix of a shunt resistance and parallel post. The current vector of the substation can be calculated according to the substation current, as in Eq. (7.37). I_{s_sub11}, I_{s_sub12} and I_{s_sub13} are the currents flowing through line 1, line 2 and line 3, respectively (Fig. 7.27).

$$Y_{sub} = \begin{bmatrix} \frac{1}{R_{sub}} & -\frac{1}{R_{sub}} & 0 \\ -\frac{1}{R_{sub}} & \frac{1}{R_{sub}} & 0 \\ 0 & 0 & 0 \end{bmatrix} + \begin{bmatrix} \frac{1}{R_{small}} & 0 & -\frac{1}{R_{small}} \\ 0 & 0 & 0 \\ -\frac{1}{R_{small}} & 0 & -\frac{1}{R_{small}} \end{bmatrix} \tag{7.36}$$

$$I_{s\_sub} = \begin{bmatrix} I_{sub} \\ -I_{sub} \\ 0 \end{bmatrix} \tag{7.37}$$

7.3.4.4 Autotransformers

The autotransformer is used in the AC railway with autotransformer configuration. As stated earlier, the admittance matrix of the whole network (Y) is a 6N × 6N matrix, therefore the admittance matrix of an element is a 6 × 6 matrix. The autotransformer model is composed of two autotransformer impedances (Z_{AT}), a magnetising impedance (Z_m), a grounding impedance (Z_{EA}) and three paralleling posts, connected to a double track system with 6 conductors (C1, C2, R1, R2, F1, F2). The magnetising impedance is very large compared to other impedance, therefore it can be ignored. The admittance matrix of an autotransformer can be calculated by Eq. (7.38). The current vector of the autotransformer can be calculated according to the autotransformer current, as in Eq. (7.39) (Fig. 7.28).

$$Y_{AT}=\begin{bmatrix}\frac{1}{z_{AT}}+\frac{1}{z_m} & -\frac{1}{z_{AT}} & -\frac{1}{z_m} & & & \\ -\frac{1}{z_{AT}} & \frac{2}{z_{AT}}+\frac{1}{z_{EA}} & -\frac{1}{z_{AT}} & 0 & 0 & 0\\ -\frac{1}{z_m} & -\frac{1}{z_{AT}} & \frac{1}{z_{AT}}+\frac{1}{z_m} & 0 & 0 & 0\\ 0 & 0 & 0 & \frac{1}{z_{AT}}+\frac{1}{z_m} & -\frac{1}{z_{AT}} & -\frac{1}{z_m}\\ 0 & 0 & 0 & -\frac{1}{z_{AT}} & \frac{2}{z_{AT}}+\frac{1}{z_{EA}} & -\frac{1}{z_{AT}}\\ 0 & 0 & 0 & -\frac{1}{z_m} & -\frac{1}{z_{AT}} & \frac{1}{z_{AT}}+\frac{1}{z_m}\end{bmatrix}+\frac{1}{R_{small}}\begin{bmatrix}1 & 0 & 0 & -1 & 0 & 0\\ 0 & 1 & 0 & 0 & -1 & 0\\ 0 & 0 & 1 & 0 & 0 & -1\\ -1 & 0 & 0 & 1 & 0 & 0\\ 0 & -1 & 0 & 0 & 1 & 0\\ 0 & 0 & -1 & 0 & 0 & 1\end{bmatrix} \tag{7.38}$$

$$I_{s\_sub}=[I_{AT}0-I_{AT}I_{AT}0-I_{AT}]^T \tag{7.39}$$

7.3.4.5 Trains

There are several equivalent circuits for different working modes of trains. The models of traction and braking trains at the up-direction track in normal operation are presented in Figs. 7.29 and 7.30.

Since the model of the trains in normal operation does not consist of a resistance, the admittance matrix is a null matrix in Eq. (7.40). The current vector depends on the direction of train current. For the traction train in up-direction in Fig. 7.29, the train collects the current from the up-tract contact line and the current returns back to lumped rails. Therefore, the current vector can be expressed in Eq. (7.41). As for the braking train in Fig. 7.30, the braking train regenerates power and feeds the contact lines. Therefore, the current vector can be expressed in Eq. (7.42).

$$Y_{train}=\begin{bmatrix}0\,0\,0\\0\,0\,0\\0\,0\,0\end{bmatrix} \tag{7.40}$$

$$I_{t\_trac\_up}=\begin{bmatrix}-I_{train}\\I_{train}\\0\end{bmatrix} \tag{7.41}$$

$$I_{t\_brake\_up}=\begin{bmatrix}I_{train}\\-I_{train}\\0\end{bmatrix} \tag{7.42}$$

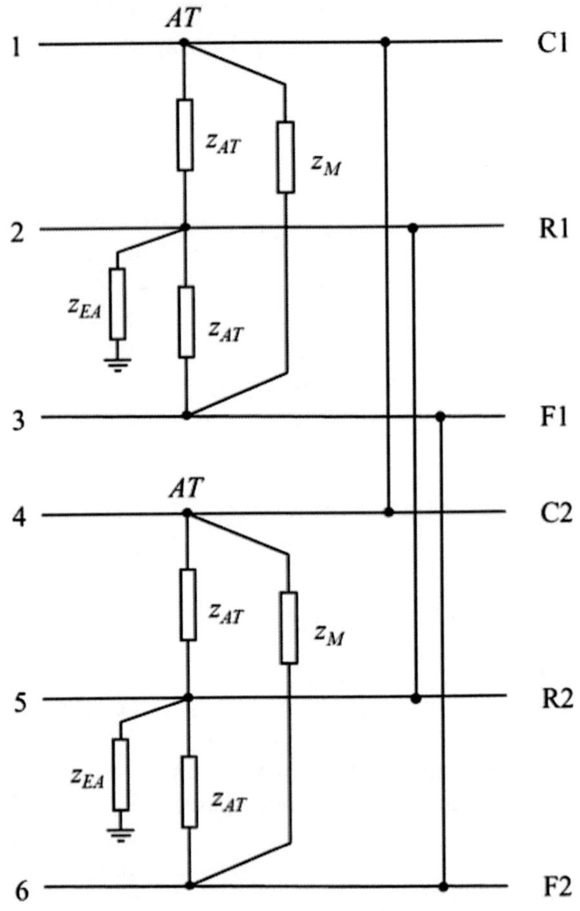

Fig. 7.28 Autotransformer element

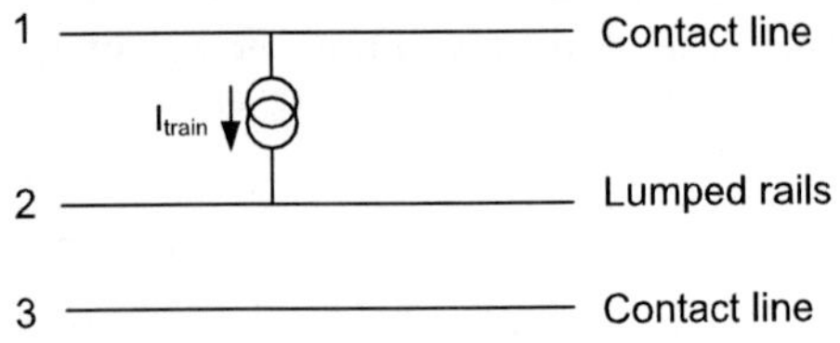

Fig. 7.29 Traction train in up direction

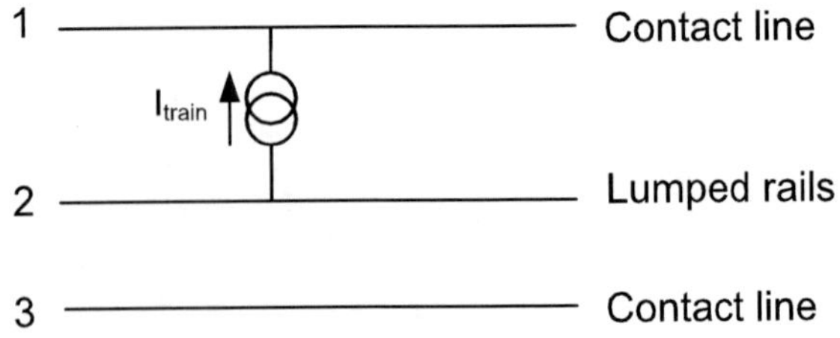

Fig. 7.30 Braking train in up direction

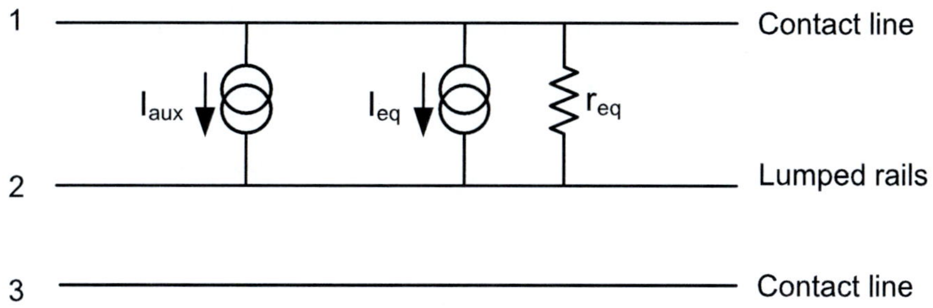

Fig. 7.31 Traction train in under-voltage operation

The model of a traction train in under-voltage operation is shown in Fig. 7.31, which can be transformed from the equivalent circuit of traction train in under-voltage in Fig. 7.16. Two current sources and a resistance are connected in parallel between the contact line and lumped rails. The admittance matrix and current vector can be expressed in Eqs. (7.43) and (7.44).

$$R_{train} = \begin{bmatrix} \frac{1}{r_{eq}} & -\frac{1}{r_{eq}} & 0 \\ -\frac{1}{r_{eq}} & \frac{1}{r_{eq}} & 0 \\ 0 & 0 & 0 \end{bmatrix} \tag{7.43}$$

$$I_{train} = \begin{bmatrix} -I_{aux} \\ I_{aux} \\ 0 \end{bmatrix} + \begin{bmatrix} -I_{eq} \\ I_{eq} \\ 0 \end{bmatrix} \tag{7.44}$$

The model of a traction train in under-voltage operation is shown in Fig. 7.32, which can be transformed from the equivalent circuit of braking train in over-voltage in Fig. 7.20. A current source and a resistance are connected in parallel between the contact line and lumped rails. The admittance matrix and current vector can be expressed in Eqs. (7.45) and (7.46).

$$R_{train} = \begin{bmatrix} \frac{1}{r_{eq}} & -\frac{1}{r_{eq}} & 0 \\ -\frac{1}{r_{eq}} & \frac{1}{r_{eq}} & 0 \\ 0 & 0 & 0 \end{bmatrix} \tag{7.45}$$

$$I_{train} = \begin{bmatrix} I_{eq} \\ -I_{eq} \\ 0 \end{bmatrix} \tag{7.46}$$

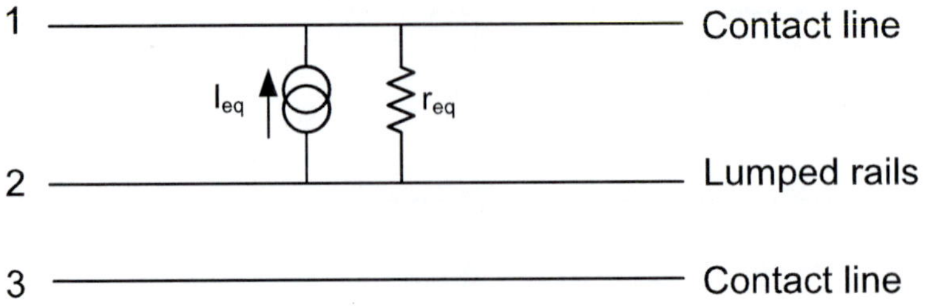

Fig. 7.32 Braking train in over-voltage operation

7.3.5 Power Flow Analysis

7.3.5.1 Current-Vector Iterative Method

When all the substations are switched on and the trains are operated in a normal working mode, the whole railway power network is a linear circuit, which consists of voltage sources (substations), current sources (trains) and resistors (transmission network). If one train is assumed as a load, the rest of the circuit becomes a linear two-terminal circuit. According to Thevenin's theorem, the linear two-terminal circuit can be replaced by an equivalent circuit comprised of an equivalent voltage source in series with an equivalent resistor [21], as shown in Fig. 7.33. The equivalent voltage is the open-circuit voltage at terminals. The equivalent resistance is the input resistance at terminals when the independent sources are turned off.

The train power demand is a known value. The train voltage can be obtained by solving Eq. (7.47). For a system with only one train, the voltage sources (substations) are connected in parallel with the same voltage. It is obvious that the equivalent voltage is equal to the no-load voltage of the substation. For a multi-train system, the equivalent voltage for one train depends on the state of the other trains. If most of the other trains are applying regenerative braking, the equivalent voltage for this train could be higher than the substation voltage. When most of the other trains are

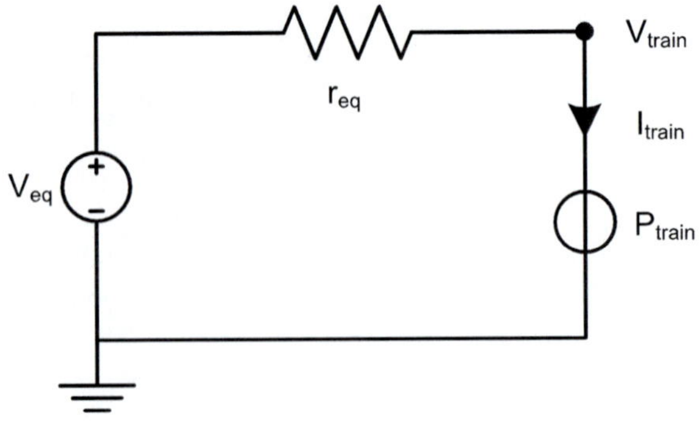

Fig. 7.33 Thevenin's equivalent circuit of a railway network

requiring traction power, the equivalent voltage for this train could be lower than the substation voltage.

$$P_{train\_demand} = \frac{(V_{eq} - V_{train})}{r_{eq}} \times V_{train} \tag{7.47}$$

If the equivalent voltage and resistance are confirmed, a quadratic formula shown as in Eq. (7.47) can be obtained. There are two theoretical solutions for a quadratic formula. The load power flow analysis aims to find the solution with higher train voltage, which is the actual train voltage in the railway network. The current-vector iterative method is presented below:

- Step 1: Initialise all the train voltage by the no-load voltage of the substation as in Eq. (7.48), where $V_{train\_n}^{(i=0)}$ is the initial train voltage.

$$V_{train\_n}^{(i=0)} = V_{sub} \tag{7.48}$$

- Step 2: Calculate the train current at the next iteration (i + 1) by Eq. (7.49).

$$I_{train\_n}^{(i+1)} = \frac{P_{train\_demand\_n}}{V_{train\_n}^{(i)}} \tag{7.49}$$

- Step 3: Update nodal voltages by the nodal analysis Eq. (7.50) using the power network admittance construction. The train voltage at this iteration can be updated at this step. The nodal analysis Eq. (7.50) is equivalent to Eq. (7.51) which is obtained from the Thevenin's equivalent circuit in Fig. 7.33.

$$\left[V^{(i)}\right] = [Y]^{-1} \times \left[I^{(i)}\right] \tag{7.50}$$

$$V_{train\_n}^{(i)} = V_{eq\_n} - r_{eq\_n} \times I_{train\_n}^{(i)} \tag{7.51}$$

- Step 4: Calculate train power at this iteration by Eq. (7.52).

$$P_{train\_n}^{(i)} = V_{train\_n}^{(i)} \times I_{train\_n}^{(i)} \tag{7.52}$$

- Step 5: Check whether the difference of calculated train power and train power demand is within the stop criterion. If so, the current-vector iterative method ends. If not, repeat step 2 to update the train current for next iteration using the calculated train voltage.

Figures 7.34 and 7.35 describe the geometrical interpretation of the current–vector iterative approach for a traction train. The parabola represents the P–V relation of Eq. (7.47). The straight green line represents the P–V relation of Eq. (7.49). The straight green line moves with the iteration. The intersection of the parabola and the straight line denotes the train power result of the iteration. The initial train voltage is the substation voltage. For common cases, the initial voltage is higher than the final result. Thus, the straight line moves anticlockwise until converging to the solution, as shown in Fig. 7.34. For some abnormal cases, the initial voltage could be lower than the final solution. The straight line moves clockwise until it converges to the solution. It is evident that the solution can be found by a current-vector iterative approach regardless of the level of initial voltage.

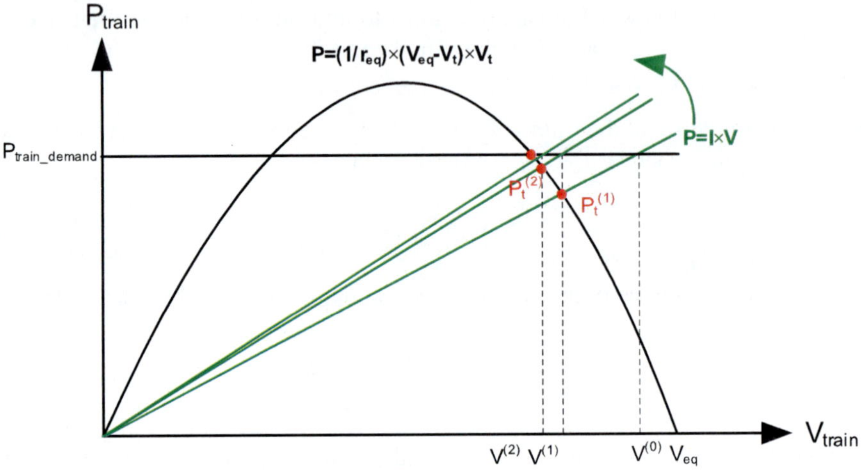

Fig. 7.34 Geometrical interpretation for a traction train

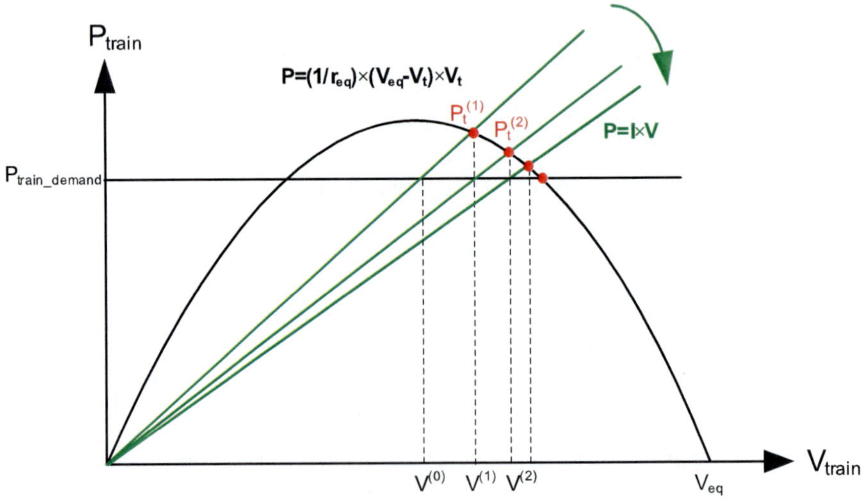

Fig. 7.35 Geometrical interpretation for a traction train in some abnormal cases

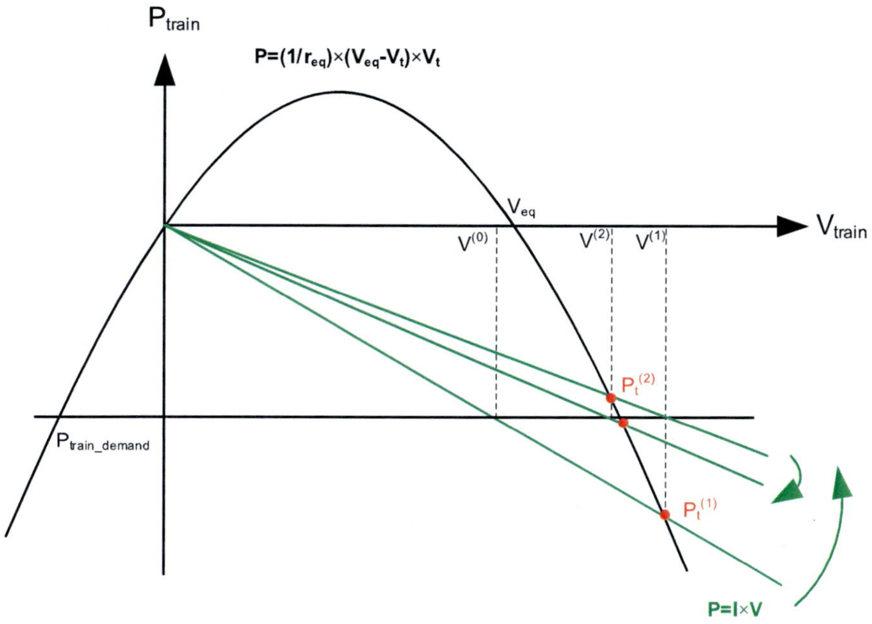

Fig. 7.36 Geometrical interpretation for a regenerative braking train

Figure 7.36 demonstrates the geometrical interpretation of the current-vector iterative approach for a regenerative braking train. The train power demand is negative for the braking train. The movement of the straight green line is always close to the solution on both sides in sequence. Whatever the initial train voltage is, the load flow solution can be found.

7.3.5.2 Working Mode Selection Algorithm

The nodal voltages can be found when the current-vector iterative method converges. Sometimes, the voltage of the train or substation mismatches the model used in the admittance matrix. Therefore, a working mode selection algorithm is required to check if the working mode matches the nodal voltage. If the current-vector iterative method converges, the following checks should be conducted:

- Step 1: Over-voltage regeneration train check: if the voltage of the normal regenerating train is higher than the highest permanent voltage V_{max1}, and the current exceeds the current limitation, change the normal regeneration train model into the over-voltage model and then rebuild the admittance matrix. If not, go to step 2;
- Step 2: Over-power regeneration train check: if the power of the over-voltage regeneration train exceeds the power limitation, change the over-voltage model

into the normal regeneration model and then rebuild the admittance matrix. If not, go to step 3;

- Step 3: Under-voltage traction train check: if the voltage of the normal traction train is lower than the under-voltage limitation $a \times V_n$, and the current exceeds the current limitation, change the normal traction train model into the under-voltage model and then rebuild the admittance matrix. If not, go to step 4;
- Step 4: Over-power traction train check: if the power of the under-voltage traction train exceeds the power limitation, change the under-voltage model into the normal traction model and then rebuild the admittance matrix. If not, go to step 5;
- Step 5: Substation voltage check: if the voltage of the substation mismatches its working mode, change the substation model to match the voltage level and then rebuild the admittance matrix. If not, the load flow analysis is correct and the power network simulation ends.

The current-vector iterative method converges in most cases. However, if the working modes of substations and train change, the current-vector iterative method may not converge to find the solutions. The geometrical interpretation of the situation when the current-vector iterative method cannot converge is shown in Fig. 7.37. This is because the train power demand is too high, which exceeds the capacity of the power network. The working mode selection algorithm is capable of detecting this situation. If the current-vector iterative method does not converge, in order to increase the capacity of the power network, all substations should be switched on. The substations are switched off due to the high regenerative power. All braking trains should be set to the over-voltage model to prevent substations from switching off.

A flow chart of the structure of the power network simulation is presented in Fig. 7.38. The power network simulator collects the network parameters and data from the motion simulation and then solves the power network by admittance matrix formulation, load flow solving and working mode selection.

7.4 Energy Flow of Railway Traction Power Systems

7.4.1 Multi-train Energy Simulator

Based on the railway system modelling methods illustrated in Sect. 7.2, a Multi-Train Energy Simulator (MTES) has been developed to evaluate the energy flow in a railway system. The structure of the MTES is shown in Fig. 7.39.

This simulator combines single-train motion simulation and multi-train power network simulation. The dynamical input parameters (driving strategies) with fixed input (traction and route data) are imported into the motion simulator. Then, the output single train trajectory and power requirement with a whole-day timetable and power network parameters are imported into the power network simulator, which will export the electrical energy consumption, including substation and transmission

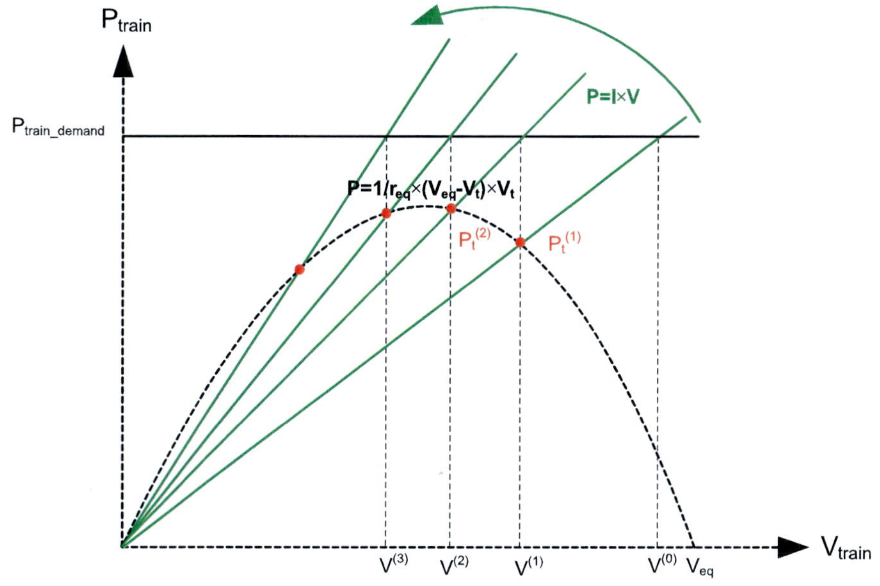

Fig. 7.37 Geometrical interpretation of divergence

losses, as well as the actual used and wasted regenerative energy. According to the energy evaluation results from the multi-train power network simulation, the dynamic inputs (driving strategies and timetable) can be modified to optimise the total energy consumption.

7.4.2 Energy Flow

In order to study the energy efficiency of the whole railway system (up to the substations), the typical energy flow diagram through the DC-fed railway is shown in Fig. 7.40. There are three layers, namely substation level, catenary system level and train level. The substations collect electricity from the national electricity grid to feed the whole railway system.

The substation energy is the bill paid by the railway operators. From the load flow analysis, the voltage and current of each substation can be obtained. The substation energy consumption is computed by integrating all substation's instantaneous power over the train operation time, as shown in Eq. (7.53).

$$E_{sub} = \int_0^T \sum_{n=1}^{N_s} \left(V_{sub\_n}(t) \times I_{sub\_n}(t)\right) dt \quad (7.53)$$

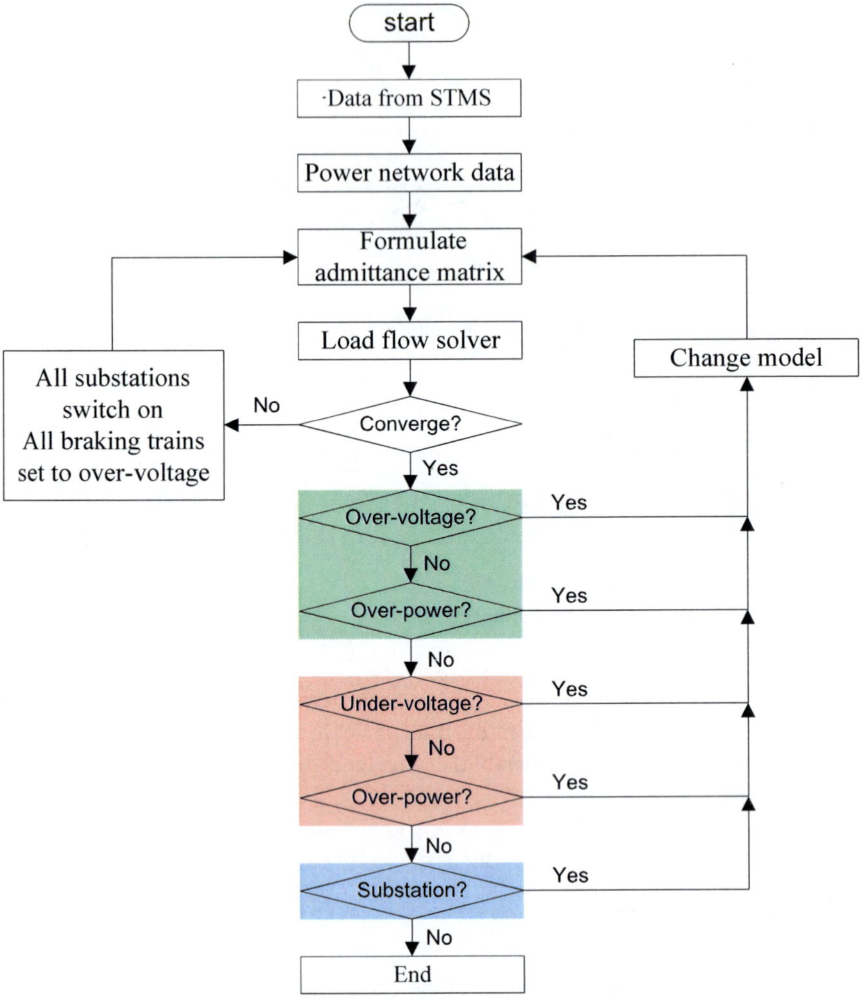

Fig. 7.38 Structure of the power network simulation

where:

- E_{sub} is the substation energy consumption;
- N_s is the number of substations;
- T is the total time of train operation;
- V_{sub_n} is the instantaneous voltage of a substation;
- I_{sub_n} is the instantaneous current of a substation.

Due to the internal resistance of substations, some energy will be dissipated inside the substations as heat. The electrical losses within each substation are determined by the losses in the transformer and diodes [22]. However, as the substation loss does

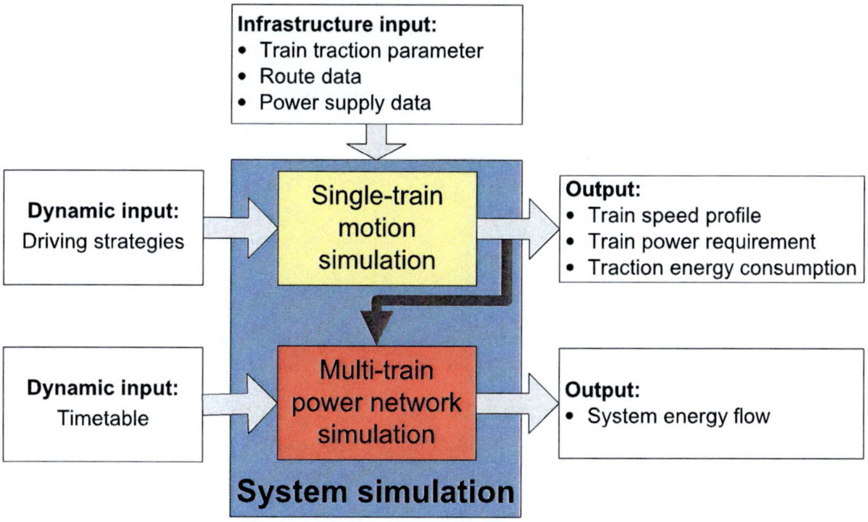

Fig. 7.39 Diagram of MTES structure

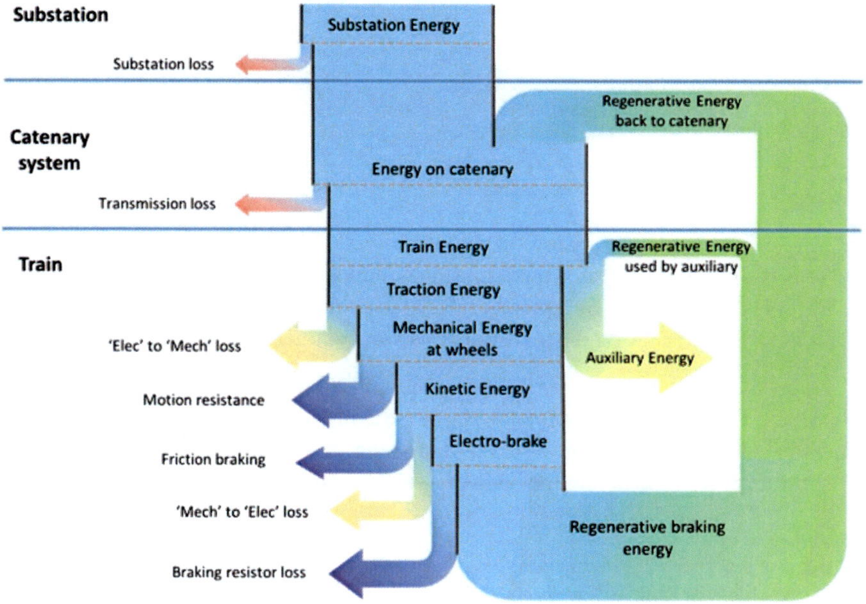

Fig. 7.40 Energy flow of DC-fed railway

not have a significant effect on the system energy evaluation and optimization, for simplicity, the substation loss is approximated using the equivalent substation inner resistance. Thus, the substation energy loss is given in Eq. (7.54).

$$E_{sub\_loss} = \int_0^T \sum_{n=1}^{N_s} \left(R_{sub} \times (I_{sub\_n}(t))^2\right) dt \tag{7.54}$$

where:

- E_{sub_loss} is the substation energy loss;
- R_{sub} is the equivalent resistance of the rectifier substation for estimating the loss.

After the losses from substations, the rest of the substation energy can be transferred to the catenary. The energy on the catenary combines some of the substation energy and the regenerative braking energy which is transferred back to the catenary system. As the current goes through the resistive transmission lines, some energy is dissipated as heat. The energy loss in transmission is given in Eq. (7.55). The resistance of the transmission conductor is a time-varying variable, which is obtained according to the train locations and network.

$$E_{trans\_loss} = \int_0^T \sum_{n=1}^{N_c} \left(R_n(t) \times (I_n(t))^2\right) dt \tag{7.55}$$

where:

- E_{trans_loss} is the transmission loss;
- N_c is the number of power transmission conductors;
- $R_n(t)$ is the resistance of a piece of transmission conductor at time t;
- $I_n(t)$ is the current of a piece of transmission conductor at time t.

Trains receive the electricity from pantographs which connect with the transmission lines. The train power depends on the voltage and current at the pantograph, which is solved by a load flow solver. Thus, the train energy can be computed by integrating all train instantaneous power over the time, as shown in Eq. (7.56).

$$E_{train} = \int_0^T \sum_{n=1}^{N_t} \left(V_{train\_n}(t) \times I_{train\_n}(t)\right) dt \tag{7.56}$$

where:

- E_{train} is the train energy consumption at the pantograph;
- N_t is the number of trains in the network;
- $V_{train\_n}(t)$ is the voltage of a traction train at time t;

- $I_{train\_n}(t)$ is the current of a traction train at time t.

The train energy and some of the regenerative braking energy are used for train traction and the auxiliary system, as shown in Eq. (7.57).

$$E_{train} + E_{regentotrain} = E_{traction} + E_{aux} \tag{7.57}$$

where:

- $E_{regentotrain}$ is the regenerative energy used by the on-board auxiliary system;
- $E_{traction}$ is train electrical traction energy consumption;
- E_{aux} is on-board auxiliary energy consumption.

The auxiliary power is assumed as constant for a train. Therefore, the auxiliary energy consumption can be calculated by integrating the power of the on-board auxiliary system over the time, shown in Eq. (7.58).

$$E_{aux} = \int_0^T \sum_{n=1}^{N_t} P_{aux}(t)dt \tag{7.58}$$

When the train is braking, a small part of the regenerative braking energy is used by the on-board auxiliary system directly. This energy can be calculated by the overlapping of auxiliary power and electric braking power, as shown in Eq. (7.59)

$$E_{regen\ to\ train} = \int_0^T \sum_{n=1}^{N_t} \min\left\{P_{aux}(t), P_{elec\_brake\_n}(t)\right\}dt \tag{7.59}$$

The train traction energy is the electricity consumed by the train for traction, which depends on driving styles. It is the sum of the mechanical energy at the wheels and the conversion loss. The mechanical energy at the wheels depends on the train tractive effort and train speed, given in Eq. (7.60).

$$E_{mech} = \int_0^T \sum_{n=1}^{N_t} \left(F_{traction\_n}(t) \times v_n(t)\right)dt \tag{7.60}$$

where:

- E_{mech} is the mechanical energy at the wheels;
- $F_{traction\_n}(t)$ is the tractive effort of a traction train at time t;
- $v_n(t)$ is the speed of a train at time t.

As discussed in the section on train motion simulation, some energy is dissipated by transforming from electrical to mechanical energy. The relation between train mechanical energy and electrical energy consumption is expressed in Eq. (7.61). In this book, the efficiency (η) refers to the whole traction chain from pantograph to wheels and is assumed as a constant. The energy loss of the conversion from electrical to mechanical energy can be calculated by Eq. (7.62).

$$E_{mech} = E_{traction} \times \eta \tag{7.61}$$

$$E_{traction\_loss} = E_{traction} \times (1 - \eta) \tag{7.62}$$

where:

- η is the efficiency of electrical to mechanical conversion;
- $E_{traction\_loss}$ is the energy loss of the conversion from electrical to mechanical.

The mechanical energy at the wheels is used to move the train and overcome the motion resistance. The energy loss on motion resistance is given in Eq. (7.63).

$$E_{motion\_loss} = \int_{0}^{T} \sum_{n=1}^{N_t} (R_n(t) \times v_n(t))dt \tag{7.63}$$

where:

- E_{motion_loss} is the energy loss on motion resistance;
- $R_n(t)$ is the motion resistance of a traction train at time t.

All the trains are assumed to be operated in a circle. Therefore, the final potential energy by gradient is zero. Thus, the rest of the mechanical energy is train kinetic energy consumption, given in Eq. (7.64). This kinetic energy is not the maximum kinetic energy obtained by the train at running, but the part of the kinetic energy which is dissipated by the braking system.

$$E_{kinetic} = E_{mech} - E_{motion\_loss} \tag{7.64}$$

where:

- $E_{kinetic}$ is the kinetic energy of trains dissipated by braking.

A blending of the electric and mechanical brake is commonly used in modern trains. Electric braking uses the traction motor as a generator to regenerate braking energy. Friction braking is used when the motor cannot provide sufficient braking effort. Therefore, part of the kinetic energy of the train is dissipated by friction braking. The rest of it is converted into electricity by electric braking with some loss of energy in the mechanical to electrical conversion, as shown in Fig. 7.40. The efficiency (η) of the mechanical to electrical conversion is assumed to be the same as the electrical to mechanical conversion in Eq. (7.61). The amount of electricity regenerated by electric braking can be expressed in Eq. (7.65).

$$E_{elec\_brake} = \left(E_{kinetic} - E_{fric\_brake}\right) \times \eta \tag{7.65}$$

$$E_{brake\_loss} = \left(E_{kinetic} - E_{fric\_brake}\right) \times (1 - \eta) \tag{7.66}$$

where:

- E_{elec_brake} is the electricity converted by electric braking;
- E_{fric_brake} is the energy dissipated by friction braking as heat;
- E_{brake_loss} is the energy loss of the conversion from mechanical to electrical.

Most trains with regenerative braking are also fitted with braking resistors in case the regenerative energy is not receptive. The regenerative braking energy equals the electric braking energy subtracted by the energy dissipated by the braking resistor, as in Eq. (7.67).

$$E_{regen} = E_{elec\_brake} - E_{r\_loss} \tag{7.67}$$

where:

- E_{regen} is the regenerated braking energy;
- E_{r_loss} is the energy dissipated by the braking resistor.

Part of the regenerated energy is used by the on-board auxiliary system directly and the rest of the regenerated energy flows back to the catenary system.

$$E_{regen} = E_{regen\ to\ train} + E_{regen\ to\ network} \tag{7.68}$$

where:

- $E_{regen\ to\ network}$ is the regenerated energy drawn back to the catenary system.

The regenerative energy fed back to the network can be computed by integrating all of the braking train's instantaneous power on the pantograph over the time, as shown in Eq. (7.69).

$$E_{regen\ to\ network} = \int_0^T \sum_{n=1}^{N_t} \left(V_{regen\_train\_n}(t) \times I_{regen\_train\_n}(t)\right) dt \tag{7.69}$$

where:

- $V_{regen\_train\_n}(t)$ is the voltage of a regenerative braking train at time t;
- $I_{regen\_train\_n}(t)$ is the current of a regenerative braking train at time t.

7.4.3 Energy Loss Analysis

The majority of the energy consumption in railway systems is used for train traction. In order to understand and improve the energy efficiency of railway systems, the energy loss through the whole network has to be studied. The power of the auxiliary system is constant and depends on the usage of electric auxiliary equipment on board. The auxiliary energy usage is not the main factor for the system energy consumption and it cannot be reduced by traction operation optimization. To simplify the study of energy evaluation and efficient regenerative braking, the auxiliary power is assumed as zero in this book. Thus, Eq. (7.68) can be simplified into Eq. (7.70). The total regenerative energy equals the regenerative energy back to the catenary system. As a result, Eq. (7.57) can be simplified into Eq. (7.71). The train energy consumption at the pantograph equals the train electrical traction energy consumption.

$$E_{regen} = E_{regen\ to\ network} \tag{7.70}$$

$$E_{train} = E_{traction} \tag{7.71}$$

As for the energy flow analysis, the energy balance equation for the whole railway network can be expressed in Eq. (7.72).

$$E_{sub} + E_{regen} = E_{train} + E_{sub\_loss} + E_{trans\_loss} \tag{7.72}$$

7.4.3.1 Energy Losses in the Network

The energy loss in the network refers to the electricity loss during the transmission from the substation to the train, which includes the substation loss and transmission loss as given in Eq. (7.73).

$$E_{network\_loss} = E_{sub\_loss} + E_{trans\_loss} \tag{7.73}$$

From Eqs. (7.54) and (7.55), it can be found that the network loss is related to contact line resistance and line current. As a consequence, the network loss principally depends on the voltage level and material of the railway system as well as the traction demand [23]. A coefficient to characterise the amount of network loss is defined in order to compare different scenarios. The 'network loss coefficient' is given in Eq. (7.74).

$$C_n = \frac{E_{network\_loss}}{E_{train}} \tag{7.74}$$

where:

- C_n is the network loss coefficient.

The network loss coefficient denotes the electricity transmission efficiency performance of the electrical infrastructure. In general, a railway system with a higher voltage level can lead to a lower network loss coefficient. The typical values for network loss coefficient can be around 18%, 14%, 8% and 6% for 600 V, 750 V, 1500 V and 3000 V DC networks, respectively [24]. The contact line resistivity is another factor of the value of the network loss coefficient. With the development of superconducting cables using high-temperature superconducting materials, the feeding loss in the power supply network could be reduced [25].

7.4.3.2 Energy Loss in Traction

Based on Eqs. (7.61) to (7.64), the traction energy equals the sum of traction conversion loss, the motion loss and the train kinetic energy, shown as in Eq. (7.75).

$$E_{traction} = E_{traction\_loss} + E_{motion\_loss} + E_{kinetic} \tag{7.75}$$

The traction conversion loss comprises 10–15% of total traction energy consumption [26]. The majority of the loss is from motor inefficiency. To reduce this loss requires more efficient motor designs, for example, selecting improved materials or using permanent magnet synchronous motors. The study of motor efficiency is not considered in this book, and the efficiency of traction conversion is assumed to be a constant. The energy used to overcome motion resistance comprises 10–50% of total traction energy. From Eq. (7.63), the amount of the motion loss depends on the motion resistance and vehicle speed. The motion resistance depends on the Davis coefficients and route curvature and gradient configuration, as discussed in the previous chapters. The method to reduce motion resistance is not the main objective of this book. The speed should be considered to reduce motion loss. The vehicle obtains kinetic energy when accelerating. Some of the kinetic energy is used to overcome the motion resistance. The rest of the kinetic energy will be converted into heat or electricity by the braking system. The amount of kinetic energy dissipated by the braking system is related to vehicle speed. This book focuses on the design of an energy-efficient driving profile to reduce the traction energy consumption.

Coasting control is one of the most significant efficient driving controls [27]. During coasting, no tractive power is applied and the acceleration is determined by the force balance of gradient and resistance. In general, the kinetic energy of the train is reduced to overcome the motion resistance. Figures 7.41 and 7.42 present the velocity and energy consumption profiles without and with coasting, respectively. The time taken by both drivings is the same, but the traction energy consumption is reduced by 17.6%. In Fig. 7.41, the train accelerates to 60 km/h and then cruises until braking. When the train is cruising, partial tractive power is applied to keep the same speed and the tractive energy increases with the increasing motion energy loss.

In Fig. 7.42, the train accelerates to a higher speed of 70 km/h and then cruises for a short period. After cruising, the train starts to coast. During coasting, no tractive power is applied and the tractive energy keeps the same. The motion energy loss still increases when the train is coasting, but this energy is covered by train kinetic energy. The train brakes until it coasts to 47 km/h.

The detailed energy comparison is shown in Table 7.2. For the same route with the same running time, driving with proper coasting controls can reduce the traction energy consumption by 17.6% (from 7.11 to 5.86 kWh). The traction loss is reduced

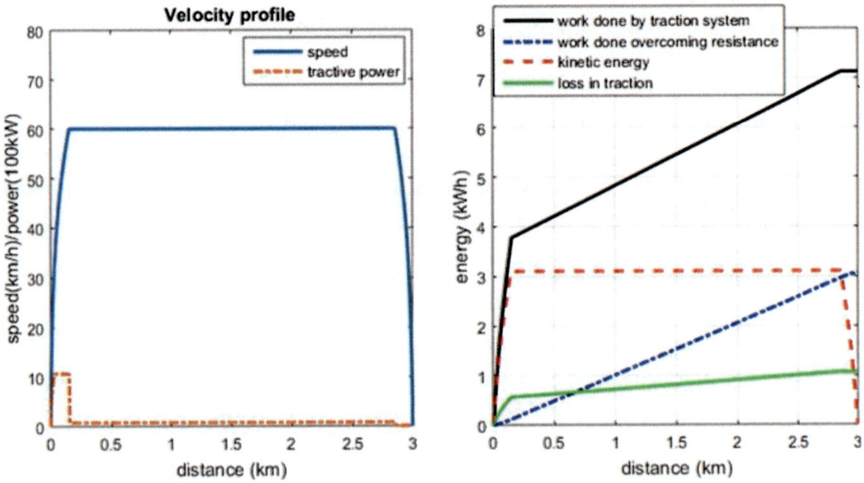

Fig. 7.41 Speed profile and traction energy case 1 (without coasting)

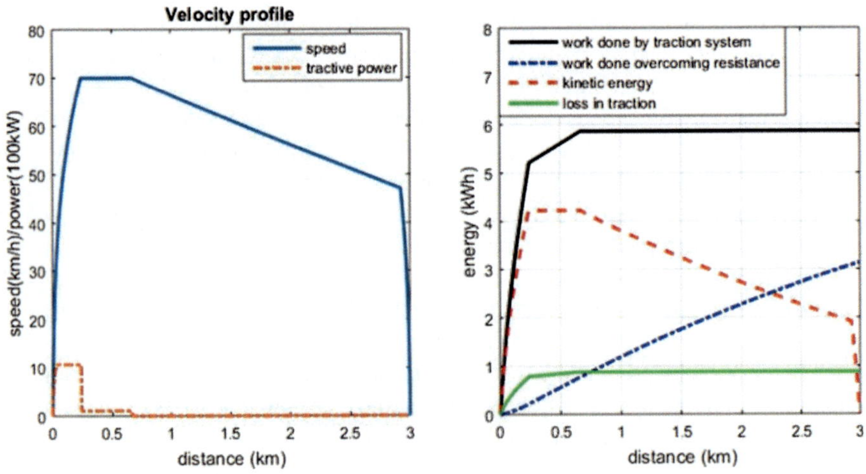

Fig. 7.42 Speed profile and traction energy case 2 (with coasting)

Table 7.2 Comparison between driving without and with coasting

Item	Driving without coasting	Driving with coasting
Distance (km)	3	3
Journey time (s)	194	194
Traction energy (kWh)	7.11	5.86 (−17.6%)
Traction loss (kWh)	1.07	0.88
Motion loss (kWh)	3.07	3.12
Kinetic energy dissipated by braking (kWh)	2.97	1.86

relatively. The motion loss for the driving with coasting is slightly higher due to the high-speed running, but the kinetic energy is reduced significantly, which is the main reason for traction energy saving. In terms of global energy optimization, if all of the kinetic energy can be reused without losses, optimising the motion energy loss is the only way to reduce the system energy consumption. Nevertheless, it is not possible due to the nature of the efficiency of conversion from kinetic energy to electricity. Therefore, traction energy consumption optimization by reducing kinetic energy is still a good solution to saving system energy for railways, although the motion loss may be increased.

7.4.3.3 Energy Loss in Regenerative Braking

Kinetic energy is converted by the braking system, and part of it is converted into electricity and reused by trains. The energy flow during braking is shown in Eq. (7.76). Since the regenerative energy can be reused, the energy loss during braking includes the friction braking energy, the electric braking conversion loss and the energy loss by electric braking resistance.

$$E_{kinetic} = E_{fric\_brake} + E_{brake\_loss} + E_{r\_loss} + E_{regen} \tag{7.76}$$

Friction braking is used when the motor cannot provide sufficient braking effort. For most metro trains, the electric braking is sufficient for normal braking requirements. Friction braking is only applied when the train is approaching a stop with very low speed. In this book, the energy dissipated by friction braking is assumed as zero, which is reasonable for a metro system energy study. Therefore, the energy flow through the braking system can be simplified, as shown in Eq. (7.77).

$$E_{kinetic} = E_{brake\_loss} + E_{r\_loss} + E_{regen} \tag{7.77}$$

The energy loss of the conversion from mechanical to electrical depends on the efficiency of the power conversion system, which is assumed as a constant in this book. Therefore, the energy dissipated by the braking resistor is the main factor to influence the amount of regenerated energy. The energy wasted in braking resistance is due to the overvoltage of the braking train. When the number of motoring trains is insufficient to absorb the regenerative energy, a high voltage will appear. This is because the train voltage can be increased by regenerative braking. In case of a high voltage hazard when the network voltage exceeds a safe value, some braking energy should be wasted in the onboard braking rheostat as heat instead of being transferred to contact lines.

In order to evaluate the efficiency of using the regenerative braking energy, the regeneration efficiency is defined in Eq. (7.78). High regeneration efficiency means a high receptivity of the network.

$$\eta_{regen} = \frac{E_{regen}}{E_{elec\_brake}} = \frac{E_{regen}}{E_{regen} + E_{r\_loss}} \tag{7.78}$$

A simple study of the regeneration efficiency has been carried out based on a 750 V railway system with a 15 km route, as shown in Fig. 7.43. There are three substations along the route. There are two trains running in the network, one traction train and one braking train.

The power flow of this network by varying the distance between these two trains is computed. The regeneration efficiency based on the braking train instantaneous power is calculated at different distances from the traction train, as shown in Fig. 7.44. When the distance between two trains is within 2 km, 100% regeneration efficiency can be achieved. The regeneration efficiency decreases when the distance increases further. Finally, the efficiency reduces to 23% when the distance is 15 km. Therefore, the distance between traction trains and braking trains is the main factor to change the efficiency of regeneration. In general, a short distance leads to high regeneration efficiency. In a railway system with busy traffic it is easier to achieve high regeneration efficiency compared with a railway system with less traffic or long headways.

From the power results shown in Fig. 7.45, the substation power increases with the train distance. The train traction power remains the same with 2000 kW. The substation power is related to the network loss and regeneration power. When the

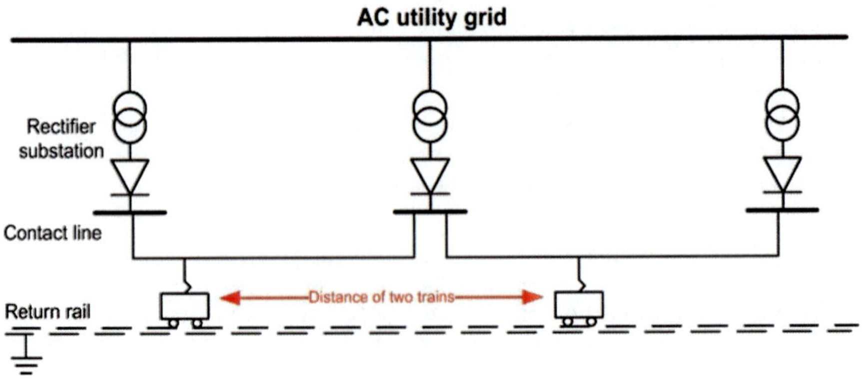

Fig. 7.43 Railway network diagram

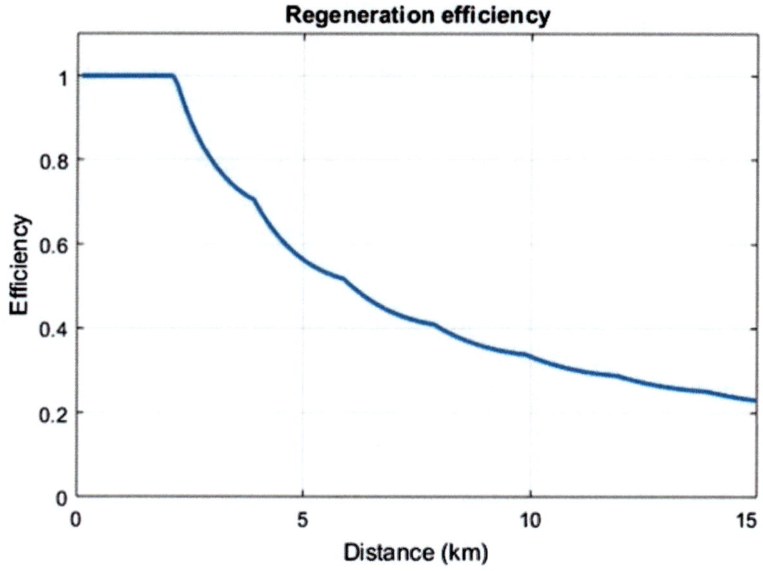

Fig. 7.44 The receptiveness of regenerative energy between two trains

distance is within 2 km, although the regeneration power is 2000 kW with 100% efficiency, the substation power varies a lot with the change in network loss. Therefore, the regeneration efficiency is significant to substation energy consumption, but the network loss cannot be neglected.

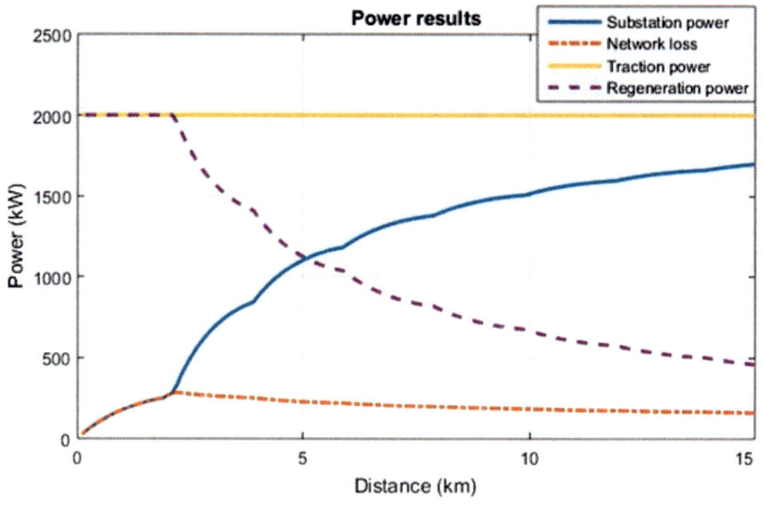

Fig. 7.45 Power results

7.5 Case Studies

The Beijing Yizhuang Subway Line (BYSL) in China has been in operation since 2010. It is a significant subway line which links the suburbs of Beijing and the city centre. Trains are operated by a mixture of Automatic Train Operation (ATO) system and human drivers. In order to assess the energy performance of the current operation of BYSL, two no-load driving tests were conducted in September 2014. One of them is the current ATO driving test, and the other is human driving with energy-efficient driving strategies. In this case study, the practical data for simulating BYSL is presented. The ATO driving speed profiles are used in this case study and the energy consumption is evaluated by the system energy simulator. The simulation results and field test results are illustrated and compared.

7.5.1 Modelling Formulation

7.5.1.1 Route Data

The Beijing Yizhuang Subway Line covers a length of 22.73 km and contains 14 stations including both underground and over ground segments. The station location data is presented in Table 7.3. A diagram of the route vertical alignment and train station locations is shown in Fig. 7.46.

Table 7.3 Station location

No	Station	Location (m)
1	Yizhuang	0
2	Ciqu	1334
3	Ciqunan	2620
4	Jinghailu	4706
5	Tongjinanlu	6971
6	Rongchang	9309
7	Rongjing	10,663
8	Wanyuan	11,943
9	Wenhuayuan	13,481
10	Yizhuangqiao	14,474
11	Jiugong	16,456
12	Xiaohongmen	18,822
13	Xiaocun	20,097
14	Songjiazhuang	22,728

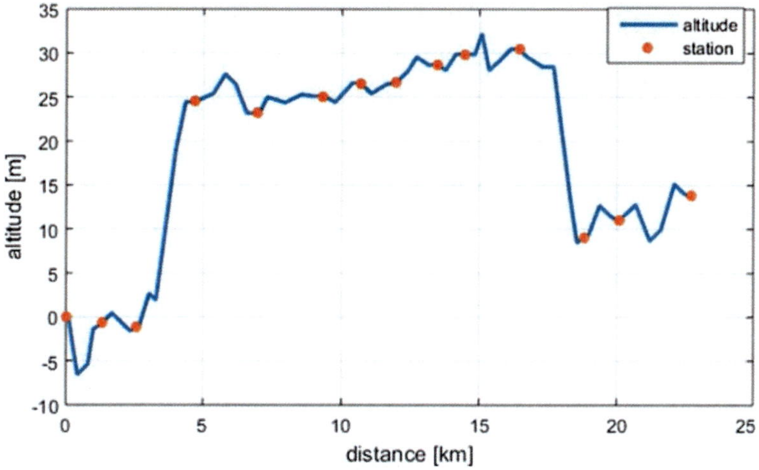

Fig. 7.46 Route vertical alignment and train station location

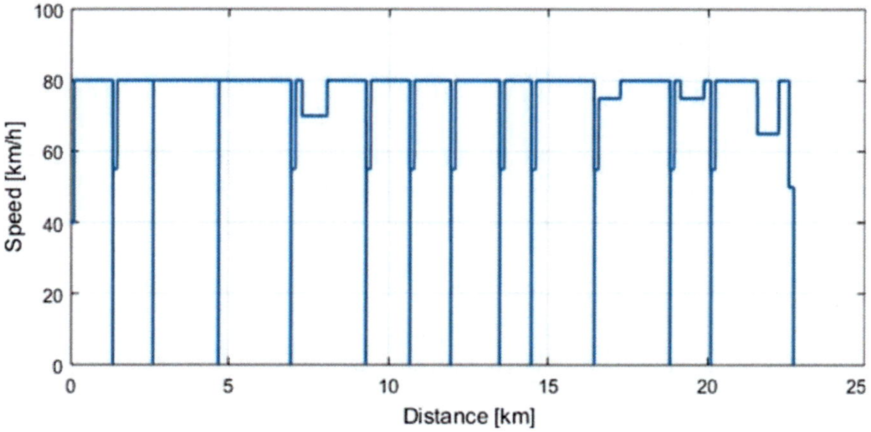

Fig. 7.47 Speed limits

The speed limits are demonstrated in Fig. 7.47. The speed limit is marked as zero at each station. The maximum speed is 80 km/h for this route. Lower speed limits are applied when the train departs from or arrives at a station.

7.5.1.2 Vehicle Data

The train operated in the Beijing Yizhuang Subway Line is formatted by 6 carriages. 3 of them are equipped with motors (M1 to M3), while the other carriages are trailers (T1 to T3). The tare weight of each carriage is shown in Table 7.4.

Table 7.4 Vehicle tare mass (tonnes)

Car No.	T1	M1	T2	M2	M3	T3	Total
Vehicle tare mass	33	35	28	35	35	33	199

Table 7.5 Passenger mass (tonnes)

Car No.	T1	M1	T2	M2	M3	T3	Total
AW0 (no load)	0	0	0	0	0	0	0
AW2 (normal load)	13.56	15.24	15.24	15.24	15.24	13.56	88.08
AW3 (over load)	17.40	19.50	19.50	19.50	19.50	17.40	112.8

Table 7.6 Tractive characteristics

	F_m (kN)	F_{m2} (kN)	V_1 (km/h)	V_2 (km/h)
AW0 (no load)	200	200	51.3	51.3
AW2 (normal load)	289	228.8	38	48
AW3 (over load)	312	228.8	35.2	48

The passenger weight of each carriage for different scenarios is presented in Table 7.5.

The tractive effort characteristic is introduced in the previous chapters. The tractive parameters for different scenarios of BYSL are shown in Table 7.6. The tractive effort curve in Fig. 7.48 describes the relationship between tractive effort and velocity in different scenarios.

The regenerative braking effort characteristic is given in Table 7.7 and the braking effort curve is described in Fig. 7.49.

The resistance to motion is calculated by Davis equation. The curvature resistance is neglected. The Davis coefficients for different scenarios are shown in Table 7.8 and the motion resistance curve is shown in Fig. 7.50.

7.5.1.3 Power Network Data

The electrical network is comprised of 12 rectifier substations with nominal 750 V supply, shown in Table 7.9. For the power network simulation, the substation is modelled as a voltage source with a no-load voltage of 850 V. The equivalent internal resistance is 0.02 Ω. The resistance rate for the overhead line and return rail are 15 μΩ/m and 10 μΩ/m, respectively.

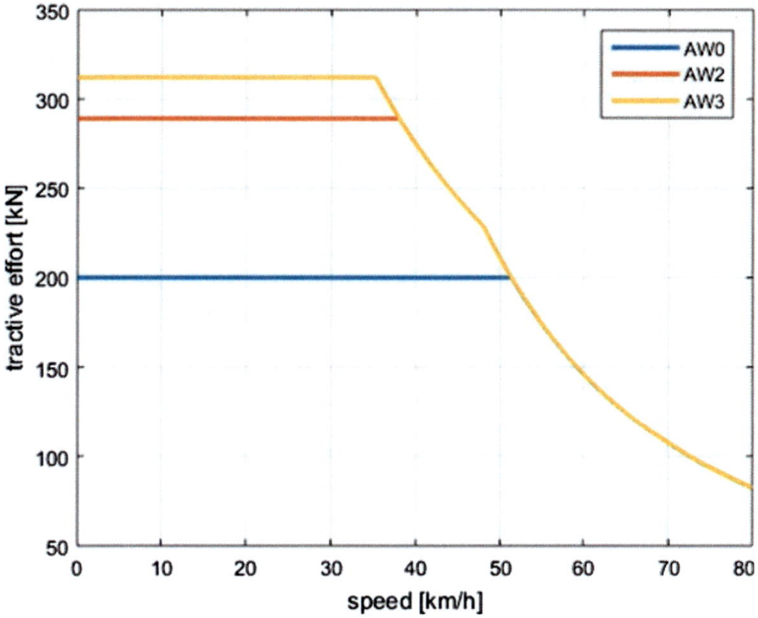

Fig. 7.48 Tractive effort curve

Table 7.7 Regenerative braking effort characteristics

	F_m (kN)	F_{m2} (kN)	V_1 (km/h)	V_2 (km/h)
AW0 (no load)	167	167	77.8	77.8
AW2 (normal load)	239	232	64	66
AW3 (over load)	255	232	60	66

7.5.1.4 Timetable

The first and last trains of BYSL depart Yizhuang Station at 5:30 and 22:05, respectively. At peak hours (from 5:30 to 9:00 and from 16:00 to 19:00), the headway is 6 min, and at off-peak hours, the headway is 11 min. There are in total 121 cycles including 65 cycles with short headway and 56 cycles with long headway. The current inter-station running time and dwell time in the train timetable is shown in Table 7.10. It is allowed to be within 5 s for each interstation running.

7.5.2 Current Driving

Trains on BYSL are driven by the ATO system and human operation jointly. In order to validate the simulation calculation and evaluate the current system energy

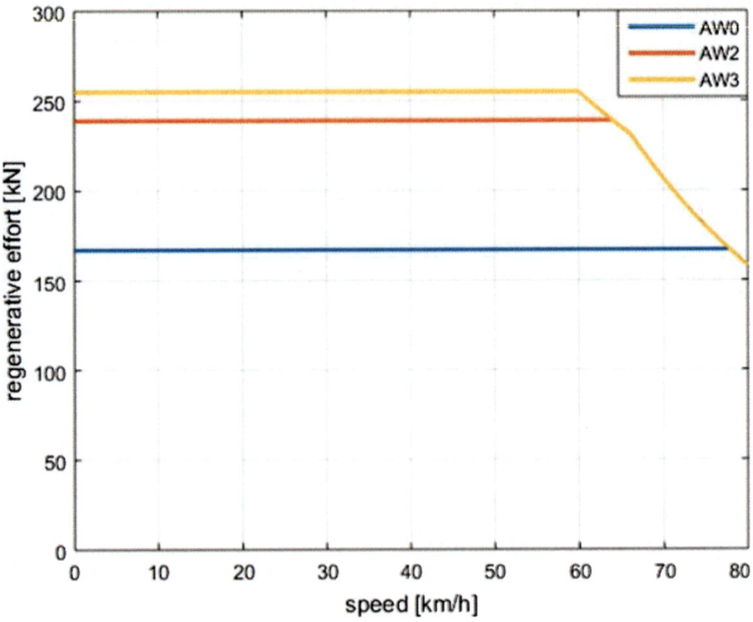

Fig. 7.49 Regenerative braking effort curve

Table 7.8 Davis constants

	A (kN)	B [kN/(km/h)]	C [kN/(km/h)^2]
AW0 (no load)	2.4180	0.0280	0.0006575
AW2 (normal load)	3.4818	0.0403	0.0006575
AW3 (over load)	3.7799	0.0437	0.0006575

consumption, the current driving data of ATO are collected by the on-board Train Information Measurement System (TIMS) of BYSL. The data includes the traction effort, power, time, speed, location and voltage. According to the collected train speed data, the train speed trajectory and mechanical power can be simulated by the Single Train Motion Simulator (STMS), shown in Figs. 7.51 and 7.52. The existing operation is to drive the train to the maximum target speed (approximately 75 km/h), and then maintain a constant speed (cruising mode) until the braking train approaches the station stop. However, due to the limitations of the ATO speed tracking algorithm and the traction characteristic, the train control is switched between motoring and braking modes frequently in order to maintain the cruising speed.

The traction energy consumption and electrical braking energy for each interstation journey are shown in Tables 7.11 and 7.12. The measured energy values come from the TIMS, and the simulated energy values are from the STMS according to the real speed trajectory. It is found that the average difference of the energy results

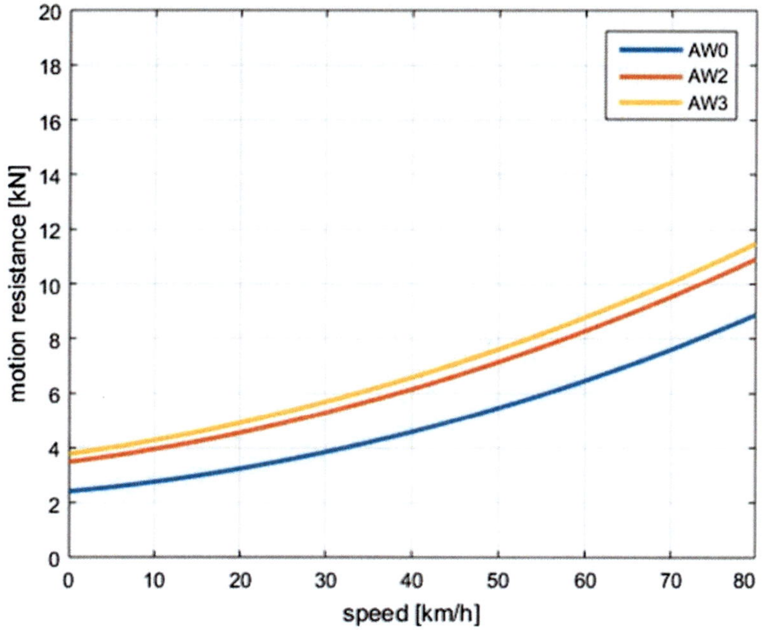

Fig. 7.50 Motion resistance curve

Table 7.9 Substation location

	Substation	Location (m)
1	Yizhuang	0
2	Ciqunan	2620
3	Jinghailu	4706
4	Tongjinanlu	6971
5	Rongchang	9309
6	Rongjing	10,663
7	Wenhuayuan	13,481
8	Yizhuangqiao	14,474
9	Jiugong	16,456
10	Xiaohongmen	18,822
11	Xiaocun	20,097
12	Songjiazhuang	22,728

between the measured and simulated values is less than 3%, which validates the accuracy of the STMS. It notes that the braking energy is much lower than traction energy between the 3rd station (Ciqunan) and 4th station (Jinghualu) due to a steep uphill. Most of the traction energy is used to cover the potential energy, and less braking is used. In addition, for the interstation journey between the 11th station

Table 7.10 Current inter-station running time and dwell time in the train timetable

	Station	Running time (s)	Dwell time (s)
1	Yizhuang	–	40
2	Ciqu	105	45
3	Ciqunan	101	35
4	Jinghailu	140	30
5	Tongjinanlu	148	30
6	Rongchang	160	30
7	Rongjing	103	30
8	Wanyuan	99	30
9	Wenhuayuan	113	30
10	Yizhuangqiao	85	35
11	Jiugong	134	30
12	Xiaohongmen	155	30
13	Xiaocun	104	30
14	Songjiazhuang	193	30
Up-direction total		**1640**	**455**
Turnaround		**–**	**180**
1	Songjiazhuang	–	30
2	Xiaocun	190	30
3	Xiaohongmen	106	30
4	Jiugong	156	30
5	Yizhuangqiao	131	35
6	Wenhuayuan	86	30
7	Wanyuan	112	30
8	Rongjing	100	30
9	Rongchang	103	30
10	Tongjinanlu	163	30
11	Jinghailu	147	30
12	Ciqunan	135	35
13	Ciqu	100	45
14	Yizhuang	103	40
Down-direction total		**1632**	**455**
Cycle total		**3272**	**1090**

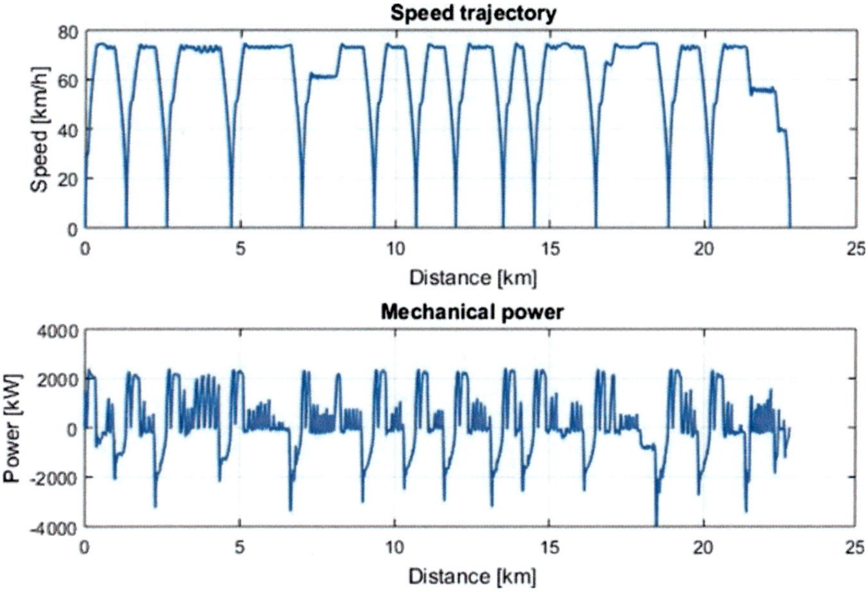

Fig. 7.51 ATO driving profiles for up-direction

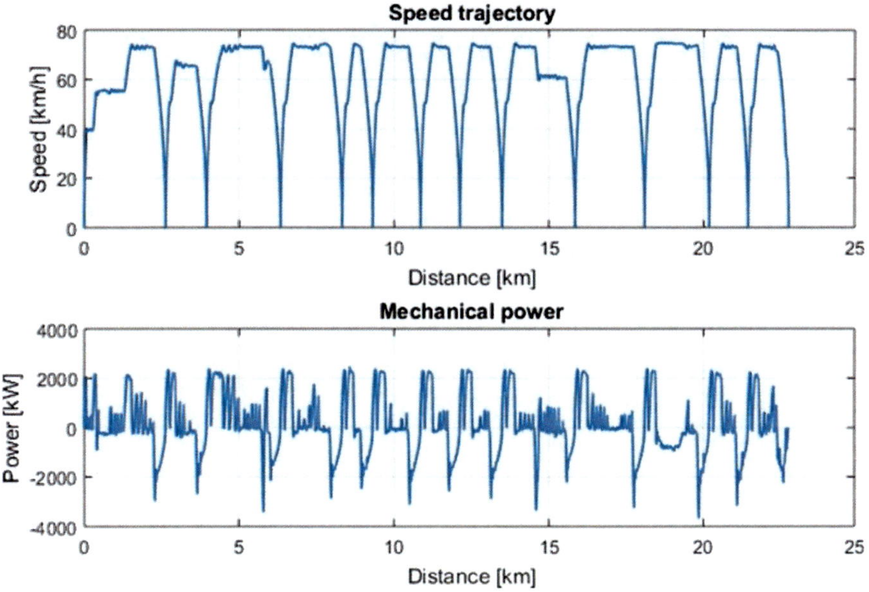

Fig. 7.52 ATO driving profiles for down-direction

Table 7.11 Train energy consumption for up-direction

	Station	Measured electrical tractive energy (kWh)	Simulated electrical tractive energy (kWh)	Measured electrical braking energy (kWh)	Simulated electrical braking energy (kWh)
1	Yizhuang	–	–	–	–
2	Ciqu	16	16.50	6	9.50
3	Ciqunan	17	18.32	11	10.78
4	Jinghailu	33	33.95	9	8.78
5	Tongjinanlu	23	21.00	12	10.97
6	Rongchang	24	22.63	13	11.21
7	Rongjing	16	18.13	10	9.64
8	Wanyuan	17	17.38	9	9.92
9	Wenhuayuan	19	20.08	12	10.39
10	Yizhuangqiao	16	17.07	10	9.67
11	Jiugong	20	20.48	12	10.70
12	Xiaohongmen	20	18.34	20	18.20
13	Xiaocun	21	20.53	10	10.72
14	Songjiazhuang	26	25.25	11	12.62
	Total	268	269.64	145	143.10

(Jiugong) and 12th station (Xiaohongmen), the braking energy is relatively high, because the corresponding route is a steep downhill.

7.5.3 *Energy Evaluation Results*

7.5.3.1 Results with Various Headways

Trains in a metro system run repetitively and periodically when the headway is constant. During the headway period, each train in a multi-train system finishes one part of the cycle running, and the sum of each train running is the whole cycle journey. Therefore, the sum of each train's traction energy during the headway period is actually the single train traction energy consumption of one cycle. The system energy evaluation for this study is always the energy consumption during the headway period rather than that of the whole day's operation time.

Figure 7.53 describes the energy results during a headway period with regeneration turned off versus different headways, which range from 240 to 900 s. The traction energy and braking energy are 524 kWh and 288 kWh, respectively, which are fixed values with different headways. The substation energy consumption ranges from 550 to 562 kWh, and there is only 2% difference with various headways. As the

Table 7.12 Train energy consumption for down-direction

	Station	Measured electrical tractive energy (kWh)	Simulated electrical tractive energy (kWh)	Measured electrical braking energy (kWh)	Simulated electrical braking energy (kWh)
1	Songjiazhuang	–	–	–	–
2	Xiaocun	26	23.34	12	13.45
3	Xiaohongmen	16	15.53	10	9.66
4	Jiugong	32	34.19	10	10.39
5	Yizhuangqiao	21	20.22	13	10.70
6	Wenhuayuan	15	15.74	10	9.78
7	Wanyuan	17	17.83	9	10.69
8	Rongjing	16	17.34	11	9.84
9	Rongchang	15	17.00	9	10.12
10	Tongjinanlu	19	21.22	11	11.74
11	Jinghailu	21	21.20	10	10.45
12	Ciqunan	12	15.25	18	18.34
13	Ciqu	20	18.88	11	10.79
14	Yizhuang	18	16.85	9	9.11
	Total	248	254.60	143	145.07

regeneration is turned off, all of the electric braking energy is dissipated by the braking resistor. The regeneration energy is zero in this case. The network loss ranges from 26 to 38 kWh, which is the reason for the substation energy changing. Although the network loss changes a lot relative to itself, it does not significantly affect the substation energy consumption.

The efficiency results are shown in Fig. 7.54. The regeneration efficiency is zero as no regeneration energy is reused. The network loss coefficient ranges from 5 to 7%, and trends to reduce with the increase of headway. This is because the average current through a network of busy traffic with low headway is higher than that through a network of quiet traffic. Therefore, the substation energy consumption tends to decrease a little with the increase of headway.

Figure 7.55 indicates the energy results with regenerating trains. As there are no changes in the train trajectory, the result of traction and braking energy are also the same as the results without regeneration. The energy consumption at the substations ranges from 291 to 446 kWh. The minimum and maximum of substation energy consumption occur when the headway is 245 s and 842 s, respectively. In principle, over 35% of the energy saving from the substation can be achieved by 'optimising' the headway. However, in practice, even small deviations in timings of a single train could result in a significant reduction (or even increase) in the effective use of regenerated power. Compared with the average substation energy consumption for the system without regeneration, the energy is reduced by 20–48% by having

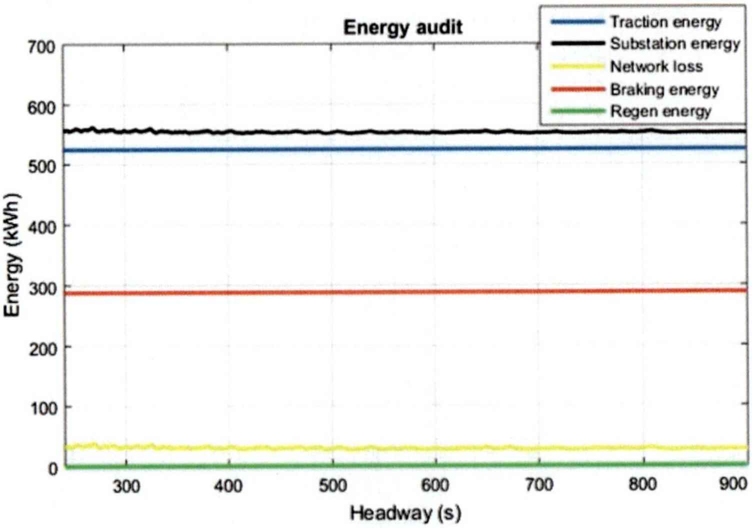

Fig. 7.53 Energy consumption with regeneration turned off

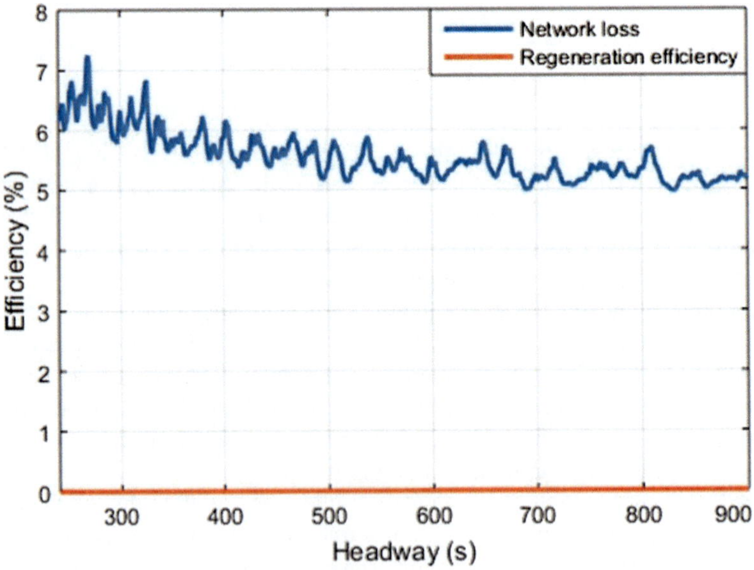

Fig. 7.54 Efficiency with regeneration turned off

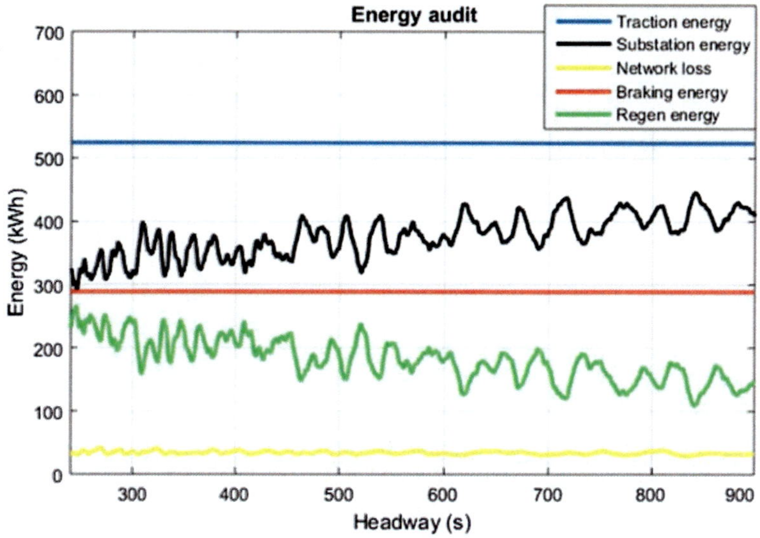

Fig. 7.55 Energy consumption with regeneration turned on

regeneration. By utilising the regeneration braking energy, at least 20% of substation energy can be saved. The regenerated energy ranges from 109 kWh (842 s headway) to 288 kWh (245 s headway), with 62% of the difference. The highly variable use of regenerated energy is not simply related to headway. This is because the station positions are at unequal distances and the braking and accelerating trains randomly overlap. However, it is clear that high regeneration energy results for short headways happen more frequently than long headways. The network losses when regeneration is turned on are a little higher, ranging from 29 to 42 kWh. But this is not significant compared to the net energy reduction.

The regeneration efficiency and network loss coefficient with different headways is shown in Fig. 7.56. The regeneration efficiency ranges from 38% when headway is 842 s to 92% when headway is 245 s. The higher regeneration efficiency can be achieved when the headway is shorter. The network loss coefficient ranges from 6 to 8%, and there is no clear trend with the increasing of headway. This is because the regenerative braking draws the current back to the network, which increases the average current through the network, even when the network is not busy.

7.5.3.2 Results After Optimization

The traction power network energy evaluation can be used to validate the energy-efficient operation strategies presented in the Chaps. 3–5. To illustrate the performance of the energy-efficient operation in the traction power network, optimization

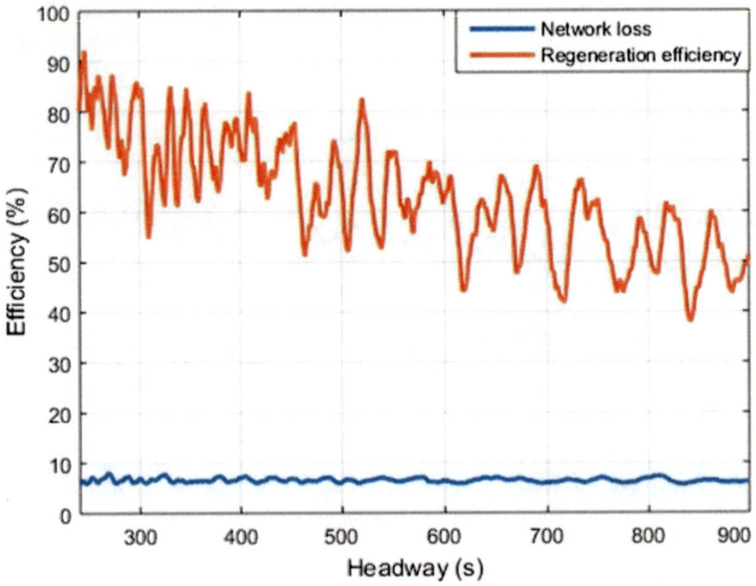

Fig. 7.56 Efficiency with regeneration turned on

results by Monte Carlo Algorithm are presented. The complete results can be found in Tian et al. [28, 29].

In this case study, the energy consumption during the peak hours is studied, where the headway is assumed as a constant with 254 s. According to the speed profile collected by the ATO system, the system energy consumption can be computed by the power network simulation. Table 7.13 presents the system energy consumption for three different operating regimes: (1) the current ATO operation, (2) the train trajectory optimization which identifies minimising the traction energy, (3) integrated train driving strategy and timetable optimization with minimising the substation energy.

The current ATO system energy consumption is calculated using a power network simulator using the speed profiles measured by the ATO system. The train trajectory optimization column in Table 7.13 presents the energy consumption of the system under traction optimization but remaining with the original timetable. The interstation journey times and dwell times are fixed and only one coasting point is used in each interstation journey. The results denote that traction energy and substation energy are reduced by 29.2% and 29.9%, respectively. With traction optimization alone, the regenerative efficiency (regenerative energy divided by braking energy) is almost the same as with ATO, at 80.6% and 82.1%, respectively. With substation energy optimization, the traction energy consumption and braking energy are almost the same as the traction energy optimization results, but the substation energy is reduced by 38.6%. This is mainly caused by the higher regenerative efficiency which reaches 95.5%.

Table 7.13 Optimization results comparison

	Current ATO operation	Train trajectory optimization	Substation energy optimization
Cycle running time (s)	4281	4281	4248
Substation energy per headway (kWh)	331.28	232.21 (*−29.9%*)	203.37 (*−38.6%*)
Substation loss per headway (kWh)	12.38	6.41	4.55
Transmission loss per headway (kWh)	26.26	16.60	16.18
Traction energy per headway (kWh)	525.94	372.52 (*−29.2%*)	375.12 (*−28.7%*)
Braking energy per headway (kWh)	289.51	199.04	201.57
Regenerative energy per headway (kWh)	233.30	163.32	192.48
Efficiency of using regenerative energy (%)	80.6	82.1	95.5

7.6 Conclusions

Railway traction power network modelling is significant to evaluate the railway system energy consumption. This chapter first illustrated the mathematical modelling methods for both DC and AC railway traction power networks. The traction power network modelling combines train operation and electrical infrastructure. The multiple trains are represented by dynamic power loads moving in the electrical power network. The power flow algorithm considering various traction modes is used to solve the electrical power network.

The traction power network model is used to analyze the energy flow for the railway system, including the energy losses in trains, traction network and regeneration. A whole system energy evaluation is studied based on the data of the Beijing Yizhuang Subway Line (BYSL). The energy consumption is calculated according to current driving by the ATO system. The energy audit result shows different system energy consumption with the regeneration braking on and off as well as the energy consumption with various headways. It has been noted that the system energy consumption with regeneration on can be reduced by 20% to 48% compared with the system with regen off, although the transmission loss is slightly increased due to higher current transmission. The system energy consumption with regeneration turned on can benefit from timetable optimization, which can, in principle, increase the efficiency of regenerative energy utilisation.

Three operation scenarios: (1) the current ATO operation, (2) the train trajectory optimization which identifies minimising the traction energy, (3) integrated train driving strategy and timetable optimization with minimising the substation energy, are presented in this chapter to illustrate the optimization results. As energy-efficient driving strategies with coasting controls are not applied in current ATO driving, speed trajectory optimization can reduce the traction energy consumption by 29.9%. The substation energy consumption is variable due to the complex interaction between the headway, the inter-station journey time, and line receptivity. Small changes in the otherwise constant headway vary the effective use of available regenerated energy significantly. The ATO driving and trajectory optimization driving have not implemented timetable optimization. Their regenerative efficiency (regenerative energy divided by braking energy) are almost very close, which are 80.6% and 82.1%, respectively. With substation energy optimization, the traction energy consumption and braking energy are almost the same as the traction energy optimization results, but the substation energy is reduced by 38.6%. This is mainly caused by the higher regenerative efficiency which reaches 95.5%.

References

1. White RD (2009) DC electrification supply system design. In: 4th IET professional development course on railway electrification infrastructure and systems, 2009. REIS 2009, pp 44–69
2. Schmid F, Goodman CJ (2009) Electric railway systems in common use. In: 4th IET professional development course on railway electrification infrastructure and systems, 2009. REIS 2009, pp 6–20
3. Baxter A (2015) Network rail—a guide to overhead electrification, p 52
4. White RD (2011) AC 25 kV 50 Hz electrification supply design. In: 5th IET professional development course on railway electrification infrastructure and systems (REIS 2011), pp 92–130
5. Kulworawanichpong T (2004) Optimising AC electric railway power flows with power electronic control. School of Engineering, Department of Electronic, Electrical and Systems Engineering, University of Birmingham, University of Birmingham
6. Fei Z, Konefal T, Armstrong R (2019) AC railway electrification systems—an EMC perspective. IEEE Electromagn Compat Mag 8(4):62–69
7. Pozzobon P (1998) Transient and steady-state short-circuit currents in rectifiers for DC traction supply. IEEE Trans Veh Technol 47(4):1390–1404
8. Goodman C (2007) Modelling and simulation. In: 2007 3rd IET professional development course on railway electrification infrastructure and systems, pp 217–230
9. Robbins AH, Miller WC (2012) Circuit analysis: theory and practice. Cengage Learning
10. Ratniyomchai T, Hillmansen S, Tricoli P (2015) Energy loss minimisation by optimal design of stationary supercapacitors for light railways. In: 2015 international conference on clean electrical power (ICCEP), pp 511–517
11. Ratniyomchai T, Hillmansen S, Tricoli P (2014) Optimal capacity and positioning of stationary supercapacitors for light rail vehicle systems. In: 2014 international symposium on power electronics, electrical drives, automation and motion (SPEEDAM), pp 807–812
12. Li X, Lo HK (2014) An energy-efficient scheduling and speed control approach for metro rail operations. Transp Res Part B Methodol 64:73–89
13. Li X, Lo HK (2014) Energy minimization in dynamic train scheduling and control for metro rail operations. Transp Res Part B Methodol 70:269–284
14. Goodman CJ, Siu LK, Ho TK (1998) A review of simulation models for railway systems. In: International conference on developments in mass transit systems (Conf. Publ. No. 453), pp 80–85

15. BS-EN50388 (2012) Railway applications—power supply and rolling stock—technical criteria. BSI
16. BS-EN50163 (2007) Railway applications—supply voltages of traction systems. BSI
17. BS-EN50641 (2014) Railway applications—fixed installations—requirements for the validation of simulation tools used for the design of traction power supply systems. BSI
18. Finlayson A, Goodman CJ, White RD (2006) Investigation into the computational techniques of power system modelling for a DC railway. In: Computers in railways X: computer system design and operation in the railway and other transit systems
19. Chymera M, Goodman CJ (2012) Overview of electric railway systems and the calculation of train performance. In: IET professional development course on electric traction systems (2012), pp 1–18
20. Mingli W, Roberts C, Hillmansen S (2010) Modelling of AC feeding systems of electric railways based on a uniform multi-conductor chain circuit topology. In: IET conference on railway traction systems (RTS 2010), pp 1–5
21. Alexander CK, Sadiku MNO (2013) Fundamentals of electric circuits, 5th edn
22. Tian Z, Hillmansen S, Roberts C, Weston P, Zhao N, Chen L, Chen M (2016) Energy evaluation of the power network of a DC railway system with regenerating trains. IET Electr Syst Transp 6(2):41–49
23. González-Gil A, Palacin R, Batty P, Powell JP (2014) A systems approach to reduce urban rail energy consumption. Energy Convers Manage 80:509–524
24. Takagi R (2010) Energy saving techniques for the power feeding network of electric railways. IEEJ Trans Electr Electron Eng 5(3):312–316
25. Takagi R (2012) Preliminary evaluation of the energy-saving effects of the introduction of superconducting cables in the power feeding network for dc electric railways using the multi-train power network simulator. Electr Syst Transp, IET 2(3):103–109
26. Douglas H, Roberts C, Hillmansen S, Schmid F (2015) An assessment of available measures to reduce traction energy use in railway networks. Energy Convers Manage 106:1149–1165
27. Tian Z, Zhao N, Hillmansen S, Roberts C, Dowens T, Kerr C (2019) SmartDrive: traction energy optimization and applications in rail systems. IEEE Trans Intell Transp Syst 20(7):2764–2773
28. Tian Z, Weston P, Zhao N, Hillmansen S, Roberts C, Chen L (2017) System energy optimisation strategies for metros with regeneration. Transp Res Part C Emerg Technol 75:120–135
29. Tian Z, Weston P, Hillmansen S, Roberts C, Zhao N (2016) System energy optimisation of metro-transit system using Monte Carlo Algorithm. In: 2016 IEEE international conference on intelligent rail transportation (ICIRT), pp 453–459

Chapter 8
Energy-Efficient Train Operation: Conclusions and Future Work

Rob M. P. Goverde, Shuai Su, and Zhongbei Tian

8.1 Conclusions

The increasing mobility, congestion and environmental impact of transport is a major concern to societies worldwide. The railways offer a sustainable mode of transport to face these societal challenges and are facing a significant growth in demand. Railway transport is the most energy-efficient means of transport due to the low friction of the steel wheel-rail interface, and specifically electric trains—representing 75% of the passenger-kilometres by rail worldwide [2]—are a sustainable means of transport, both in terms of energy efficiency and greenhouse gas emissions. The increasing railway transport demand and saturating railway capacity motivate further research and deployment of energy-efficient train operation. Train headways are getting much shorter by modern signalling and control technology, as well as planning for optimal usage of infrastructure to accommodate the growing train frequencies. This gives additional challenges to maintain stable railway traffic without unnecessary braking and re-accelerating, while also opportunities arise to re-use regenerative braking energy by nearby trains.

Train drivers can use a variety of driving strategies to operate a train from stop to stop and as such use the available running time supplements in different ways. Energy-efficient train operation makes use of optimal cruising speeds (speed holding)

R. M. P. Goverde (✉)
Department of Transport and Planning, Delft University of Technology, Delft, The Netherlands
e-mail: R.M.P.Goverde@tudelft.nl

S. Su
State Key Laboratory of Rail Traffic Control and Safety, Beijing Jiaotong University, Beijing, China
e-mail: shuaisu@bjtu.edu.cn

Z. Tian
School of Engineering, University of Birmingham, Birmingham, UK
e-mail: Z.Tian@bham.ac.uk

S. Su et al., *Energy-Efficient Train Operation*, Lecture Notes in Mobility,
https://doi.org/10.1007/978-3-031-34656-9_8

and coasting (no traction) to save energy, whilst accelerating as fast as possible to reach the optimal cruising or coasting speed, and braking as fast as possible before speed restrictions and stops. The optimal sequence and switching points between these four energy-efficient driving regimes—maximum acceleration, cruising by partial traction, coasting, maximum braking—depend on the available running time supplement, the rolling stock characteristics, and the track characteristics such as gradients, curves, tunnels and speed limits that may vary along a route. For short distances, the cruising regime may be absent when the optimal cruising speed cannot be reached before coasting already has to start. When regenerative braking can be applied another cruising regime by partial regenerative braking on declines can also be used. On steep inclines (where maximum tractive effort is not sufficient to maintain speed) and steep declines (where speed increases while coasting), the optimal switching point to respectively maximum acceleration and coasting is earlier than the incline/decline and it also lasts longer before switching back to cruising after the slope.

The optimal train trajectories can be derived using optimal control theory, and in particular by application of Pontryagin's Maximum Principle. The resulting train trajectories can be visualized in time-distance and speed-distance diagrams, and likewise the associated optimal control and energy consumption can be visualized in diagrams as function of distance or time. The main difficulty in the train trajectory optimization problems is the determination of the switching points between driving regimes, which can be many depending on the track characteristics, especially the gradient profile and speed limit profile.

Energy-efficient train trajectories should be incorporated in the railway timetabling process. Traditionally, timetable planning is based on calculations of the technical minimum running time between two stops to which a running time supplement is added as a fixed percentage or an absolute number, after which the result is rounded up to the required precision (mostly minutes). As a next step, the optimal train trajectory can be computed for the given fixed scheduled running time. This train trajectory gives the optimal drivable time-distance train path that can be used to derive target passage times at critical intermediate timing points, as well as the optimal cruising speed and coasting points for punctual driving. This detailed timetable information can be provided to the drivers to assist in energy-efficient driving as opposed to merely providing the scheduled departure and arrival times at the stops. In particular, this gives essential timing information for possible intermediate short stops and passage times at railway junctions, as well as speed advice for speed restrictions and before steep slopes such as tunnels and bridges.

Any track occupation conflict should be solved in a conflict-free timetable as a basis for energy-efficient train operation that avoids unnecessary braking and re-acceleration due to route conflicts. Therefore, microscopic railway timetabling based on blocking time theory is replacing traditional normative macroscopic timetabling in order to optimally allocate train paths to scarce and saturated railway infrastructure. The blocking time theory guarantees that signalling constraints are incorporated correctly in the timetable design. The microscopic infrastructure occupation associated to a train trajectory can be derived in terms of blocking times, which is used for conflict detection between train paths by which conflict-free timetables can be obtained. Train trajectory optimization algorithms should be used as the main running time calculation method, such that the resulting timetables and infrastructure occupation are based on energy-efficient train speed profiles. This will guarantee a perfect alignment with the actual train operation, in particular when supported by Driver Advisory Systems (DAS) or Automatic Train Operation (ATO).

Regenerative braking energy may contribute to the total energy consumption. The related strategies include synchronization of arrivals and departures in the train timetable, coordinated train control between braking and accelerating trains, and the application of energy storage systems, which should all be considered to realize the optimal usage of regenerative energy. In particular, coordinated control between regenerative braking trains and nearby accelerating trains with reused regenerative energy via the power supply network has potential for energy savings. Information of train operation and the active power substation can be shared to generate a more comprehensive energy management strategy in real-time such that the total energy consumption is minimized. Hence, this requires an integrated systems approach between train operation and power supply system.

On-board and track-side energy storage systems can be implemented in the railway systems to improve the utilisation of the surplus regenerative braking energy and reduce traction power demand. The original energy-efficient train driving strategies are no longer the optimal strategies for railway systems with energy storage devices. New objective functions considering energy storage models are proposed for the optimization. Some case studies based on the Metro in Madrid have been developed considering various operation and infrastructure scenarios. Track-side energy storage systems reduced energy consumption by 9.72–10.19% in scenarios with low-density traffic. However, the energy reduction ratio decreased to 1.8–2.07% in scenarios with high-density traffic. On the other hand, installing on-board energy storage systems could be beneficial in low-traffic lines with low receptivity to regenerated energy, where savings between 7.46% and 11.67% have been obtained. However, their application in dense traffic conditions or with high network receptivity would increase energy consumption because of the increase in train mass.

Electric trains are supplied with energy by the traction power supply network. Multiple trains are moving along the power supply network requiring traction power demand when accelerating and feeding back the regenerative braking power when decelerating. Modelling of the multi-train traction power supply network plays a significant role in evaluating railway system energy flow and validating the energy-efficient train operations. Both DC and AC traction power supply networks with various voltage levels are widely used across the world. The equivalent circuits of substations, transmission network and trains can be derived by mathematical equations. The power flow through the traction power supply network and multiple trains can be solved by piecewise iterative power flow analysis algorithms. The train speed trajectory and timetabling models provide the real-time train power demand and location, which can be fed into the power network simulation to calculate the detailed energy flow including the substation energy consumption, transmission losses, and regenerative braking energy utilisation. The multi-train power network simulation can be used to validate, evaluate and optimize the energy-efficient train operation strategies.

8.2 Future Work

Timetables are the basis for the real-time traffic plan that a railway traffic management system maintains to match the real-time route setting to the train paths [6]. The real-time traffic plan contains the allocation of successive track sections and blocks to train paths. In particular, in a digitalized railway system the real-time traffic plan can be used by DAS and ATO trackside systems to compute train path envelopes (or journey profiles) that are sent to the DAS/ATO onboard systems [9]. These train path envelopes define the target times or time windows at timing points along the route of a train that can be used as constraints by an onboard train trajectory optimization algorithm to compute the energy-efficient train trajectory given the actual rolling stock parameters and conditions, while avoiding conflicts with other train trajectories. In practice, the parameter values and conditions may differ from the ones assumed during timetabling, and therefore the actual train trajectories may also deviate more or less from the planned ones. For instance, in a periodic railway timetable the scheduled running times will be the same for all trains associated to a given train line that run with a given frequency over the day. However, the train compositions may vary over the day with different train lengths corresponding to the fluctuating transport demand over different periods of the day. Hence, these train compositions vary in length, mass, resistance and traction characteristics, and therefore the individual optimal train trajectories may also vary slightly. In addition, a traffic management system monitors train delays and adjusts the timetable in case of disturbances to maintain a conflict-free traffic plan using conflict detection and resolution algorithms. The updated real-time traffic plan will then provide new targets to the trains via updated train path envelopes. Then the task of the driver or the DAS/ATO onboard systems is to generate and track train trajectories within the provided train path envelopes, such that the trains operate energy efficiently in green waves over the network. The interaction between the traffic management systems, DAS/ATO trackside and DAS/ATO onboard systems, as well as the signalling systems for mainline railways is an active research area.

A recent research area is cooperative train control of a convoy of trains that move synchronously as close as possible. Train separation by more than the absolute braking distance has always been a safety principle in railways to avoid collisions in the case a train might derail. A major paradigm shift is the virtual coupling concept where successive trains use vehicle-to-vehicle communication to virtual couple and proceed as one platoon following a master train with coordinated traction and braking [7]. In this case the trains can follow at a relative braking distance while keeping a safety margin to the rear of the predecessor even when this predecessor executes emergency braking [8]. Under virtual coupling, the dynamics of the trains in a platoon will be dominated by tracking the relative distance and speed with the predecessor and the master train rather than tracking a dedicated train trajectory. This will affect the energy consumption of the platoon. The optimal joint multi-train trajectory optimization of a platoon regarding energy efficient driving of all virtually-coupled trains is still an open research topic.

In recent years, the digital twin technology (DT) has become one of the most popular research directions with high expectations. The digital twin is meant as the virtual and digital counterpart of a physical system that can be used to simulate the real system for various purposes, exploiting a real-time synchronization of the sensed data coming from the field [5]. With the vigorous development of the Internet-of-Things, high-speed communication, big data analysis, intelligent algorithms and other technologies, the digital twin technology has sufficient support from theory to technology, and is gradually applied to various fields such as industry and medicine. Different from traditional simulation, the advanced digital twin technology enables the model to have significant advantages such as real-time interaction, data-driven, and independent adjustments.

As an important part of the railway system, the traction power supply system is responsible for providing energy and power for running trains. Energy flow in the traction network is an important characteristic for the traction power supply system operation, and is the main basis for system capacity configuration and energy saving optimization. Constrained by the complexity of the system and security factors, modelling the energy flow for a traction power supply system is generally realized by traditional model simulation methods [10]. A data-driven modelling method based on the digital twin technology can be adopted to accurately simulate the energy flow for (urban) rail traction power supply systems in the future. Combined with the general architecture of a digital twin model [1], the digital twin architecture of an urban rail traction power supply system can be designed, including six major components: a physical layer, perception layer, transmission layer, data layer, computing layer and application layer. The physical layer refers to the physical object of the digital twin model, i.e., the real physical entity of the traction power supply system. Based on sensing and data acquisition technology, the perception layer collects a variety of quantities of the state representing the electrical characteristics of the physical layer, and then uploads the collected data to the data layer server at a high speed through the transmission layer. The data layer stores the original collected data to the cloud database, and processes the data. As the core part of the architecture, the computing layer improves the accuracy of the model. The computing layer mainly includes the integrated system model, correction algorithm, simulation model and the other key content. The application layer is the decision-making and design optimization application based on the model results and measured data. The decision-making optimization results are stored in the data layer and can directly act on the physical layer for decision-making guidance.

To achieve net-zero railway systems, the study of interactions of the railway network with the electrical power grid and renewable energy generation is attractive. A smart soft open point allows a controllable energy exchange between railway network and power grid to transfer the surplus power from the railway network to support the weak power grid [3]. Renewable energy and energy storage hubs can be installed along the railway network to increase the railway power supply capacity with lower grid upgrading costs and reduce the peak power demand. It also provides an opportunity in developing railway self-sufficient energy supply networks [4]. The main challenge is to control the large-scale railway and energy systems in a coordinated and smart way, which requires the development of digital and control technologies. With the advantages of the digital twin based modelling technology for the railway traction power supply system, real-time information interactive transmission is enabled between the digital model and the real-world system. The real system can be perceived in real time according to the collected state data, and meet the data requirements for the model itself. Therefore, digital twin based technology will enable the management of trains with renewable energy flow, promoting sophisticated integrated energy-saving methods of energy management and train operation.

References

1. Bao J, Guo D, Li J, Zhang J (2019) The modelling and operations for the digital twin in the context of manufacturing. Enterprise Information Systems 13(4):534–556
2. IEA (2019). *The Future of Rail: Opportunities for energy and the environment*. International Energy Agency, https://www.iea.org/reports/the-future-of-rail
3. Kamel T, Tian Z, Zangiabadi M, Wade N, Pickert V, Tricoli P (2022) Smart soft open point to synergically improve the energy efficiencies of the interconnected electrical railways with the low voltage grids. Int J Electr Power Energy Syst 142:108288
4. Kano N, Tian Z, Chinomi N, Hillmansen S (2022) Comparison of renewable integration schemes for AC railway power supply system. IET Electrical Systems in Transportation 12(3):209–222
5. Negri E, Fumagalli L, Macchi M (2017) A review of the roles of digital twin in CPS-based production systems. Procedia manufacturing 11:939–948
6. Quaglietta E, Pellegrini P, Goverde RMP, Albrecht T, Jaekel B, Marlière G, Rodriguez J, Dollevoet T, Ambrogio B, Carcasole D, Giaroli M, Nicholson G (2016) The ON-TIME real-time railway traffic management framework: A proof-of-concept using a scalable standardised data communication architecture. Transportation Research Part C: Emerging Technologies 63:23–50
7. Quaglietta E, Wang M, Goverde RMP (2020) A multi-state train-following model for the analysis of virtual coupling railway operations. Journal of Rail Transport Planning & Management 15:100195
8. Quaglietta E, Spartalis P, Wang M, Goverde RMP, van Koningsbruggen P (2022) Modelling and analysis of Virtual Coupling with dynamic safety margin considering risk factors in railway operations. Journal of Rail Transport Planning & Management 22:100313
9. Wang Z, Quaglietta E, Bartholomeus MGP, Goverde RMP (2022) Assessment of architectures for Automatic Train Operation driving functions. Journal of Rail Transport Planning & Management 24(100352):2022
10. Zhang G, Tian Z, Tricoli P, Hillmansen S, Wang Y, Liu Z (2019) Inverter operating characteristics optimization for DC traction power supply systems. IEEE Trans Veh Technol 68(4):3400–3410